Skills
Observing, Measuring, Recording, Experimenting, Cooperating, Classifying, Collecting and Analyzing Data, Noticing and Articulating Patterns, Predicting, Inferring, Drawing Conclusions

Concepts
Substances, Properties, Chemical Composition, Evaporation, Air Currents, Pressure, Surface Tension, Patterns, Polygons, Polyhedrons, Angles, Light, Color, Minimal Surface Area, Diameter, Radius, Volume

Science Themes
Scale, Structure, Matter, Energy, Stability, Patterns of Change, Systems & Interactions, Models & Simulations, Diversity & Unity

Mathematics Strands
Measurement, Geometry, Patterns and Functions

Nature of Science and Mathematics
Cooperative Efforts, Interdisciplinary, Real-Life Applications, Theory-Based and Testable

by
Jacqueline Barber
Carolyn Willard

Great Explorations in Math and Science (GEMS)
Lawrence Hall of Science
University of California at Berkeley

One class posted a sign on their classroom door during the *Bubble Festival* that said, "Having Fun—Do Not Enter." Several teachers could not resist the temptation. They came in to see what was going on, and before they knew it, kids were teaching them how to blow bubbles and demonstrating their discoveries.

"In reading my students' journals, I was amazed to find how important bubble activities were to making them feel at home with the other students in the class and with my classroom in general. What a great activity to do very early in the year."

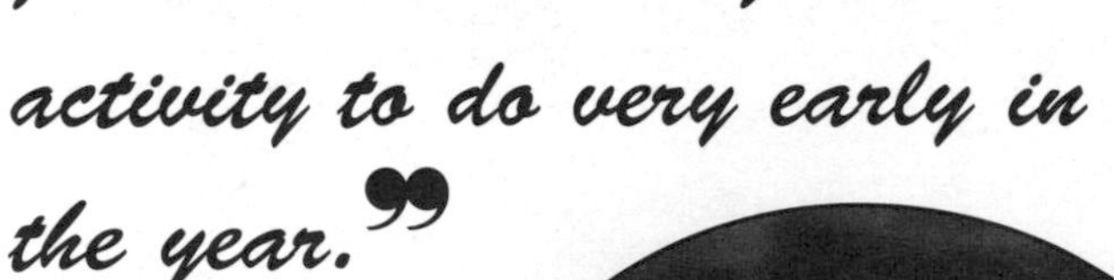

One teacher allowed her students to discover for themselves that bubbles hate dry things. This was an exciting discovery for her students and became one of the outcomes of the first exploration session.

"One first grade student in my class told me how she wished she could do things like this at home, but her family's home life is so busy, that there's never enough time for things like this. I'm convinced that allowing kids a chance to explore at their own pace is filling a deep need for a generation of kids growing up in a fast-paced world."

One group of students was working with animals in the science lab around the time they participated in a *Bubble Festival.* They were able to connect different bubble shapes they saw with structures of the animals: the compound eye of the crayfish and the fly, cells from lizard skin that they examined under the microscope. Some students tied in bubble shapes to the structure of stomates that are found on the undersides of leaves. Many of the students compared the bubble shapes they saw with the structure of geodesic domes that are common in our area.

Contents

contents

Learning Station Activities

with Activity Task Cards for Volunteers

More Elaborate Learning Station Activities

with Activity Task Cards for Volunteers

Illustrations
Carol Bevilacqua
Lisa Klofkorn
Carolyn Willard

Photographs
Richard Hoyt

Lawrence Hall of Science
Chairman: Glenn T. Seaborg
Director: Marian C. Diamond

Publication of *Bubble Festival* was made possible by a grant from the people at Chevron USA .

The GEMS project and the Lawrence Hall of Science greatly appreciate this support.

Initial support for the origination and publication of the GEMS series was provided by the A.W. Mellon Foundation and the Carnegie Corporation of New York. GEMS has also received support from the McDonnell-Douglas Foundation and the McDonnell-Douglas Employees Community Fund, the Hewlett Packard Company, and the people at Chevron USA. GEMS also gratefully acknowledges the contribution of word processing equipment from Apple Computer, Inc. This support does not imply responsibility for statements or views expressed in publications of the GEMS program. Under a grant from the National Science Foundation, GEMS Leader's Workshops have been held across the country. For further information on GEMS leadership opportunities, or to receive a publications brochure and the *GEMS Network News*, please contact GEMS at the address and phone number below.

International Standard Book Number: 0-912511-80-X

COMMENTS WELCOME

Great Explorations in Math and Science (GEMS) is an ongoing curriculum development project. GEMS guides are revised periodically, to incorporate teacher comments and new approaches. We welcome your criticisms, suggestions, helpful hints, and any anecdotes about your experience presenting GEMS activities. Your suggestions will be reviewed each time a GEMS guide is revised. Please send your comments to:
GEMS Revisions,
Lawrence Hall of Science,
University of California, Berkeley, CA 94720.

Our phone number is (510) 642-7771.

Great Explorations in Math and Science (GEMS) Program

The Lawrence Hall of Science (LHS) is a public science center on the University of California at Berkeley campus. LHS offers a full program of activities for the public, including workshops and classes, exhibits, films, lectures, and special events. LHS is also a center for teacher education and curriculum research and development.

Over the years, LHS staff have developed a multitude of activities, assembly programs, classes, and interactive exhibits. These programs have proven to be successful at the Hall and should be useful to schools, other science centers, museums, and community groups. A number of these guided-discovery activities have been published under the Great Explorations in Math and Science (GEMS) title, after an extensive refinement process that includes classroom testing of trial versions, modifications to ensure the use of easy-to-obtain materials, and carefully written and edited step-by-step instructions and background information to allow presentation by teachers without special background in mathematics or science.

Staff

Principal Investigator
Glenn T. Seaborg
Director
Jacqueline Barber
Curriculum Specialist
Cary Sneider
Staff Development Specialists
Katharine Barrett, John Erickson, Jaine Kopp, Kimi Hosoume, Laura Lowell, Linda Lipner, Laura Tucker, Carolyn Willard
Mathematics Consultant
Jan M. Goodman
Administrative Coordinator
Cynthia Ashley
Distribution Coordinator
Gabriela Solomon
Art Director
Lisa Haderlie Baker
Artists
Carol Bevilacqua and Lisa Klofkorn
Principal Editor
Lincoln Bergman
Senior Editor and Designer
Carl Babcock

Contributing Authors

Jacqueline Barber
Katharine Barrett
Lincoln Bergman
Jaine Kopp
Linda Lipner
Laura Lowell
Linda De Lucchi
Jean Echols
Jan M. Goodman
Alan Gould
Kimi Hosoume
Sue Jagoda
Larry Malone
Cary I. Sneider
Jennifer Meux White
Carolyn Willard

Acknowledgments

Several of these bubble learning stations—created by Jacqueline Barber and Nancee Boice—premiered at the "First Ever Bubble Festival" held at the Exploratorium in San Francisco in 1983. These learning stations were subsequently taken to county and urban fairs, and to various other community and school celebrations. General Electric Foundation funds brought bubble learning stations to multi-generational events held at senior citizen centers in the San Francisco Bay Area.

In 1987 the staff of the Chemistry Education program at the Lawrence Hall of Science developed a *Bubble Festival* that could serve entire elementary schools of kindergarten through sixth graders. This major effort, funded by the William K. Holt Foundation, involved many people, including Richard Silberg, Jan Coonrod, Leigh Agler, Jacqueline Barber, and Cynthia Ashley, and resulted in an extremely popular and successful series of programs. Since then, our all-school *Bubble Festival* has been further evolved by every staff member who presents it, including Laura Lowell, Kevin Beals, Sylvia Velasquez, Mayumi Shinohara, Randall Fastabend, and D. Lance Marsh. The *Bubble Festival* still goes out to schools, thanks to the people at Chevron USA.

This learning station guide, which adapts the all-school event for individual classrooms, has involved the help and hard work of many people. A list of trial teachers appears in front of this book. Special mention must be accorded to Ann Quinlan-Miani, whose extensive experience in successfully using learning stations in her classroom led to some major innovations in our stations. Co-author Carolyn Willard also designed all of the drawings for the signs. The "Frozen Bubbles" station is an adaptation of an exhibit designed by Ilan Chabay of the New Curiosity Shop in Mountain View, California. The use of stirrers and pipe cleaners for bubble skeletons was inspired by a "Family Science" building activity. Kevin Beals and Laura Lowell (veteran *Bubble Festival* presenters) and Lincoln Bergman, the GEMS Principal Editor, joined the two authors of this guide in poring over every detail of teacher feedback and reviewing each of the many drafts of this guide. Lincoln also wrote the bubble poetry and helped put together the background material that appears in the "Behind the Scenes" section. Carl Babcock provided his editing, design, and desktop publishing expertise in producing both the national trial version and this edition. Victor Ichioka assisted in adapting the sign drawings to proper computer format.

Last but far from least, thanks to all the parent volunteers and teaching assistants who endured (and enjoyed!) the enthusiastic bubbling over of their students and helped measure how much youthful excitement can fit in a room!

Reviewers

We would like to thank the following educators who reviewed, tested, or coordinated the reviewing of this series of GEMS materials in manuscript and draft form. Their critical comments and recommendations, based on presentation of these activities in classrooms nationwide, contributed significantly to these GEMS publications. Their participation in the review process does not necessarily imply endorsement of the GEMS program or responsibility for statements or views expressed in these publications. Their role is an invaluable one, and their feedback is carefully recorded and integrated as appropriate into the publications. Thank You!

ARIZONA
Coordinator: Richard Clark

Manzanita Elementary School, Phoenix
- Linda K. Carter
- Terry Dalton
- Deborah S. Miller
- Sandy Stanley

CALIFORNIA

Huntington Beach GEMS Center
Coordinator: Susan Spoeneman

Agnes L. Smith School, Huntington Beach
- Mary Dyer
- Margaret E. Hayes
- Janis D. Liss
- Judy Miskanic

College View Elementary School, Huntington Beach
- Marshall Baldwin
- Martha Deal
- Gretchen McKay
- Linda Koch Mosier
- Kathy O'Steen

Loara and Marshall Elementary Schools, Anaheim
- Judy Swank

Palm Lane and Jefferson Elementary Schools, Anaheim
- Peggy Okimoto

San Francisco Bay Area
Coordinator: Cynthia Ashley

Columbus Intermediate School, Berkeley
- Susan DeWitt
- Jose Franco
- Katherine Lunine
- Richard Silberg

Glassbrooke Elementary School, Hayward
- Marianne Camp
- Mary Davis

Henderson Elementary School, Benicia
- Cathy Larripa
- Sheila Ruhl
- Carol A. Pilling
- Barry Wofsy

Hoover Elementary School, Oakland
- Margaret Coleman
- Wanda Price

Oxford Elementary School, Berkeley
- Anita Baker
- Joseph Brulenski
- Carol Bennett-Simmons
- Sharon Kelly
- Judy Kono
- Janet Levinson

Park Day School, Oakland
- Karen Corzan
- Joan Wright-Albertini

Steffan Manor School, Vallejo
Don Baer
Sheila Himes
Annie Howard
Anna Reid
Bob Schneider
Monica A. Spini
Randy Stava

Stoneman Elementary School, Pittsburg
Varan Garro
Ann Quinlan-Miani
Linda Pineda
Joan Shelton

Travis Elementary School, Vallejo
Joyce Corcoran
Janet Matthews

Windrush School, El Cerrito
Margo Lillig
Barbara Minton
Louise Perry
Hillary Smith
Martha Vlahos

GEORGIA
Coordinator: Judy Dennison

Seaborn Lee Elementary School, Atlanta
Linda Keller
Kay Likins
Cindy Riley
Christine Ann Walker

KENTUCKY
Coordinator: Dee Moore

Dunn Elementary School, Louisville
Claudia George
Pam Kleine-Kracht
Cynthia Macshmeyer
Judith R. Stubbs

Coleridge-Taylor Elementary School, Louisville
Nancy Benge
Vicki Bumann
Andrea Lentz
Ruth Ann Little

NEW YORK
Coordinator: Stan Wegrzynowski

D'Youville Porter Campus School, Buffalo
Barbara Johnson
Marikay Reville Loftus
Lynette Parker
Deborah Ann Rudyk

Science Magnet School #59, Buffalo
Sue M. Combs
Janet E. Lawrence
Sharon Pikul
Gail M. Russo

OREGON
Coordinators: Ann Kennedy and Karen Levine

Cedaroak Park Elementary School, West Linn
Dennis E. Adams
Sue Foster
Jaylin Redden-Hefty
Cheri Weaver

TEXAS
Coordinator: Karen Ostlund

Deepwood Elementary School, Round Rock
Lana Culver
Sara Field
Myra P. Fowler
Karen Lena
M. Mercado
Sherry Wilkison

WASHINGTON
Coordinators:
Scott Stowell and Diana D'Aboy

Arlington Elementary School, Spokane
Esther C. Baker
Pauline Eggleston
Linda Graham
JoAnn Lamb
Sue Rodman
Dorothy Schultheis
Becky Sherwood
Gayle Vaughn

Bemiss Elementary School, Spokane
Ethel Barstow
Lee Anne Fowler
Bettie Maron
Cathy Mitchell

For many years the Lawrence Hall of Science has presented *Bubble Festivals* as large group events, with up to 150 people at a time participating. An entire school can take part in one or two days. These extremely popular festivals are presented during the school day for all students, in the evening and on weekends in a family setting, or at community events.

The demand for a way to adapt *Bubble Festival* activities to good advantage in the classroom setting was so great that we began working with classroom teachers and experimenting with a learning station format. This guide is the result of that effort.

Bubble Festival differs from most other GEMS teacher's guides in that it presents students with a variety of different challenges at learning stations set up around the classroom. A "learning station" is a classroom table, cluster of desks, or counter-top, set up with equipment or materials designed to encourage students to make their own discoveries. The class participates in open-ended exploration in an informal, student-centered way.

As with all GEMS guides, we welcome your comments and suggestions for changes in upcoming editions.

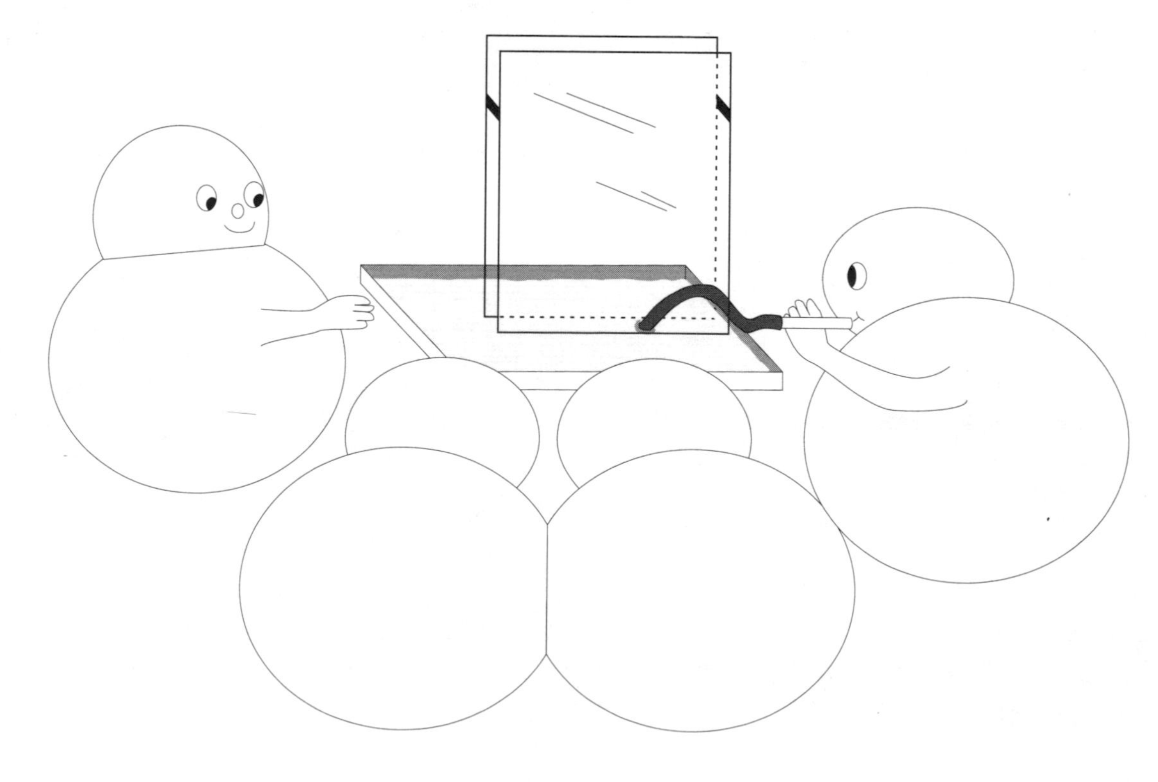

Introduction

A Program of Joyous and Merry Activity

At the "heart" of the *Bubble Festival* is student-centered excitement and open-ended investigation. In whatever ways you approach it (and this guide provides many possible variations based on actual classroom experience) there can be no doubt that you and your students will share a festive experience, in the best sense of the word.

Imagine a classroom full of students going eagerly from station to station to try a variety of fascinating experiments with bubbles. Signs at each different bubble station challenge students to explore a variety of concepts, but the student's own sense of wonder and curiosity is paramount.

Visualize:

- A student who seems timid about science and math becomes engrossed in an activity, and then realizes she is doing (and enjoying!) what a scientist does.
- A student with below-average academic skills figures out a new way to measure the height of a bubble, and enjoys an inestimably important sense of success.
- All of your students intensely involved in what they are doing, gaining through "play" an intuitive sense of what science is all about, cooperating as needed, while learning at their own pace and level.

After this festival of bubbles, picture yourself extending some of the concepts the students discovered into your regular curriculum. Perhaps your students gained firsthand experience measuring the circumference, radius, and volume of bubbles at the "Bubble Measurement" station; you might reinforce these concepts in your mathematics program. Or you may decide to teach more about the polygons they discovered at the "Shapes" or the "Skeletons" station. In your science class, you may extend students' experiences with surface tension or light and color. Writing about bubbles is another wonderful extension of a *Bubble Festival*, whether it be a journal-style record of what was discovered and what new questions arose, or the basis for fantasy writing or poetry. See "Writing and Bubbles" on page 135 and "Children's Literature Connections to a Bubble Festival" on page 136.

Great Expectations From Great Explorations

It's important at the start to have realistic and flexible expectations about what students may gain from this unit. It is our experience that there are many more beneficial educational outcomes from these activities than may at first meet the eye. We are convinced that the learning experience provided by both the content and the process of the *Bubble Festival* is multi-layered and can be extremely valuable.

Students learn at a level that is appropriate to them. Even though a certain station might offer an opportunity to learn about angles, *pi*, or surface tension, that does not mean a first-grader should be expected to assimilate these ideas. The free exploration of a *Bubble Festival* allows children to learn some important things that may be difficult for us to anticipate, label, or even to detect at the time. In "Bubble Measurement," for example, the approaches of your students will depend on background and experience. While one student may focus on the concept that a bubble *can* be measured, another may notice a relationship between circumference and diameter. Keeping the open-ended tone to all the activities will help enable all students to succeed. In "Bubble Shapes," for younger students seeing and describing the various shapes is valuable. Even for older students, while it may be satisfying to hear students "discover" or name geometric shapes, keep in mind the tremendous value of their own open-ended shape explorations. Some of the most valuable learning may be occurring when we think our students are "just having fun." Lessons learned, or intuitions gained, through direct practical experience, may have much greater impact than words alone can provide. A first grader thus may internalize experiences that will stand him in very good stead when he learns a more advanced concept in later grades. A sixth grader may be able to reinforce her intellectual description of surface tension with a tactile memory that is worth the proverbial 1,000 words.

At a glance, the idea of a *Bubble Festival* may appear to some as "too much fun and games" and not enough structured and specific content. Other teachers might assume that because their students' efforts may seem somewhat unfocused during portions of the activities, students will not walk away with increased knowledge of the many important scientific concepts and principles involved. Still other teachers may feel that while some older students may grasp these concepts, younger children will not.

It has been our consistent experience, over years of fostering the guided discovery approach to science and mathematics education, that activities such as the ones presented in the *Bubble Festival* learning stations, which involve students in open-ended investigation, are of great value at every grade level, even when they not connected to a more structured or goal-oriented curriculum. In turn, this free investigation, which is central to all scientific endeavor, provides a solid foundation for more structured learning and for guiding students toward further investigation.

During the *Bubble Festival* refrain from attempting to get the class or any individual student "back on task." Instead, observe what your students are investigating, whatever it may be. Then, by asking focused questions, you can help students articulate a discovery and/or add challenges to help students continue down a specific learning path.

Making Connections

For older students, and for teachers who are using the *Bubble Festival* guide to explore major themes that run throughout the sciences, it is important to make clear that the free experimentation of the festival by no means precludes discussion and investigation of themes such as:

- **Matter** (as seen in the nature of soap films)
- **Energy** (minimal surface area)
- **Stability** (properties common and inherent to soap films, predictable qualities)

Patterns of Change (directing student attention to color changes, or to how the shape of a bubble is changed by its proximity to other bubbles, etc.)
Structure (of bubbles themselves, in clusters, in skeletons)

There are many other examples. It turns out that bubbles are not only among the most playful phenomena in the world, their study also reveals much about numerous fields in science and mathematics. Seeing these larger connections is exciting. See the "Behind the Bubble" section on page 127 for additional notes on the larger scientific significance of bubbles.

The GEMS teacher's guide *Bubble-ology* introduces and explores many of these important concepts in the physical sciences. *Bubble-ology* is a more structured unit designed for use in Grades 5–10, as contrasted with the more elementary level of the *Bubble Festival* and its more informal learning station approach. There are also a number of stations in the *Bubble Festival* that are not part of *Bubble-ology*. Even where some overlap exists, each guide provides the student with a different and unique learning experience.

For example, "Bubble Colors" in *Bubble Festival*, in which children discover different colors and patterns, can lead nicely to "Predict-a-Pop" in *Bubble-ology*, in which students record the sequence of colors they observe and make generalizations about their observations. Similarly, in the "Bubble Measurement" festival station children discover that bubbles can be measured. In *Bubble-ology*, students use measurement as part of a controlled experiment to compare different brands of soap solution.

If you choose to use both guides, keep in mind that for older students, the *Bubble Festival* could provide an excellent introduction to the more formal, structured, guided discovery activities of *Bubble-ology*.

What About the Mess?

Let's just burst right out with the truth, which comes out in the wash anyway! Bubble activities, and the free experimentation of this learning station format, do tend to be quite messy. Bubble solution can be very slippery, spills of it are not uncommon, and there is a certain amount of chaos in even the most disciplined classroom when the magic of "bubble stuff" is unleashed. You need to be aware of this before you begin.

However, we strongly suggest that it would be a big mistake not to allow students to try these activities solely because of this kind of inconvenience. There are numerous and effective ways to minimize the mess, protect the classroom, encourage student involvement, help control and facilitate logistics—and end up with a clean classroom.

If potential mess is a matter of great concern to you, please read the suggestions on page 36 before drawing any final conclusions. We are convinced that the educational worth and positive group energy of *Bubble Festival* far outweigh the temporary inconvenience. **Nearly every classroom teacher who helped us test the *Bubble Festival* was enthusiastic about presenting it again!**

Where Science and Enjoyment Meet

This guide is not organized by class period or session, nor written to suggest a recommended teaching sequence. It is designed to be flexible. The following sections of this book contain useful general information. Descriptions of specific Learning Station Activities begin on page 54.

A teacher may choose to present any number of the activities in this guide, and can combine activities in a variety of ways. We've included teachers' suggestions for successful ways to use these learning stations in a classroom situation, and ideas for ways to sequence their presentations. **You are best situated to anticipate the needs and interests of your students, and to select those stations that best complement the other activities and subject matter you've planned for the class. GEMS wants to hear about your approach and about your students' reactions!**

We've also included ways to extend student's experiences at each learning station with "Going Further" activities, many of them multi-disciplinary in nature. Toward the end of this guide is a section on how to present a *Bubble Festival* to a larger group of 100 to 200 people. Also provided are scientific background, information resources, and children's literature connections.

Even though bubbles have no sharp edges, we like to think of *Bubble Festival* learning station activities as being on the "cutting edge" of education and enjoyment. Certainly the bubble is one place where joy and science meet, as in these *Bubble Festival* couplets:

Bubble domes and walls of bubble,
Pop them single, blow them double,

At stations take a turn to learn,
Curiosity's foam begins to churn.

Asking questions, seeking why,
Bubbles drifting in the sky,

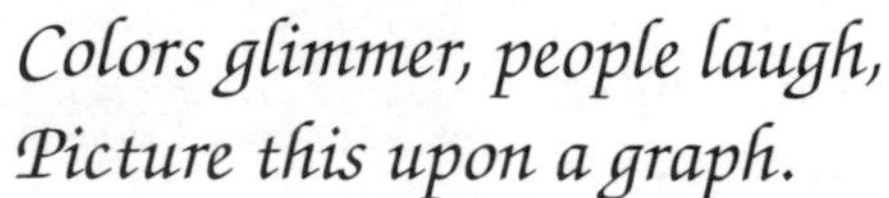

Colors glimmer, people laugh,
Picture this upon a graph.

Soap solution of pure pleasure—
Shapes and sizes we can measure.

Could our new idea be testable?
Find out at the Bubble Festival!

Reflections of a Fellow Teacher

It is worth quoting one 4th grade teacher, who described her *Bubble Festival* experiences with the following thoughts:

"I felt great anxiety before the *Bubble Festival*, with all the preparation of signs and materials, and anticipation of how wild my students might become. I felt worried and more than a little bit stressed! However, to my surprise and delight, my students were 100% involved, on-task, and cooperative. I was amazed at how smoothly all sessions of our *Bubble Festival* went and how much my students got out of it. Well yes, the clean-up was substantial, but somehow in hindsight, it all seems worth it.

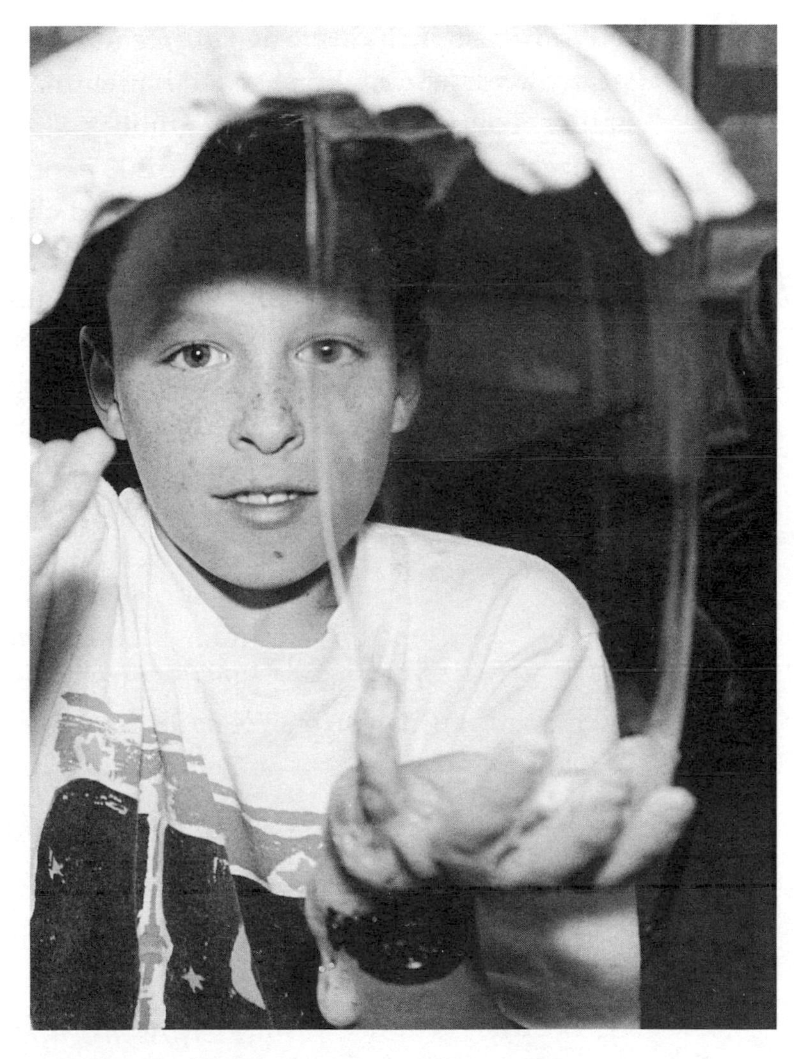

I am convinced that the experience of *Bubble Festival* fulfilled some deep needs within my students for open-ended exploration and unfettered wonderment. The discoveries my students made keep popping up in their journals, and in other activities.

For several students, *Bubble Festival* was the thing that won them over to learning in my classroom. My advice to others: this is an ambitious undertaking with incredible returns. If it piques your interest at all, try it! If you're nervous about it, you're normal! Indeed, there will be a high level of noise, excitement, and activity—brace yourself for that! But by setting clear behavioral expectations for your students, you can ensure a safe and productive setting. Get together some adult volunteers to help. If you're still worried, consider pairing up with another teacher or two to present it.

Bubble Festival is not for everyone, but it sure was for me!

I'd do it again in a minute."

Planning a Classroom Festival of Bubbles

Choosing the Format of Your *Bubble Festival*, and Which Learning Station Activities to Present

The first step in planning your *Bubble Festival* is to decide how many sessions you will present, how you would organize each session, and what specific learning station activities you want to include.

How Many Sessions?

Many people have successfully presented *Bubble Festival*s at community events, on a back-to-school night, during a family math and science day, or to the entire 4th and 5th grade of a school at one time. (Notes on presenting *Bubble Festival*s to large groups appear in the last section of this guide.) In these situations, it is most common for a *Bubble Festival* to be a one-time event, because of the nature of these situations, and the effort it takes to put on a large group event.

In contrast, individual teachers have usually set up learning station activities in their classrooms for their students to explore in a series of sessions. Most teachers found that about three to five sessions are a good number to enable students to get past the initial excitement of the bubbles and explore phenomena in more depth, while still sustaining their high interest.

How Long Should Each Session Be?

Most teachers have found that 40-60 minutes per session is appropriate, with the first sessions being somewhat longer than the last sessions. However, there are notable exceptions where teachers preferred to allow their students to explore for a longer period of time, and a few teachers who found shorter sessions easier to manage.

How Many Learning Stations Should There Be?

In most classrooms, six learning stations per session are appropriate, enabling between 4–6 students to work at each. You have the option of selecting all of the same learning station, all different learning stations, or any combination in between. The advantages and disadvantages of each of these formats are discussed below.

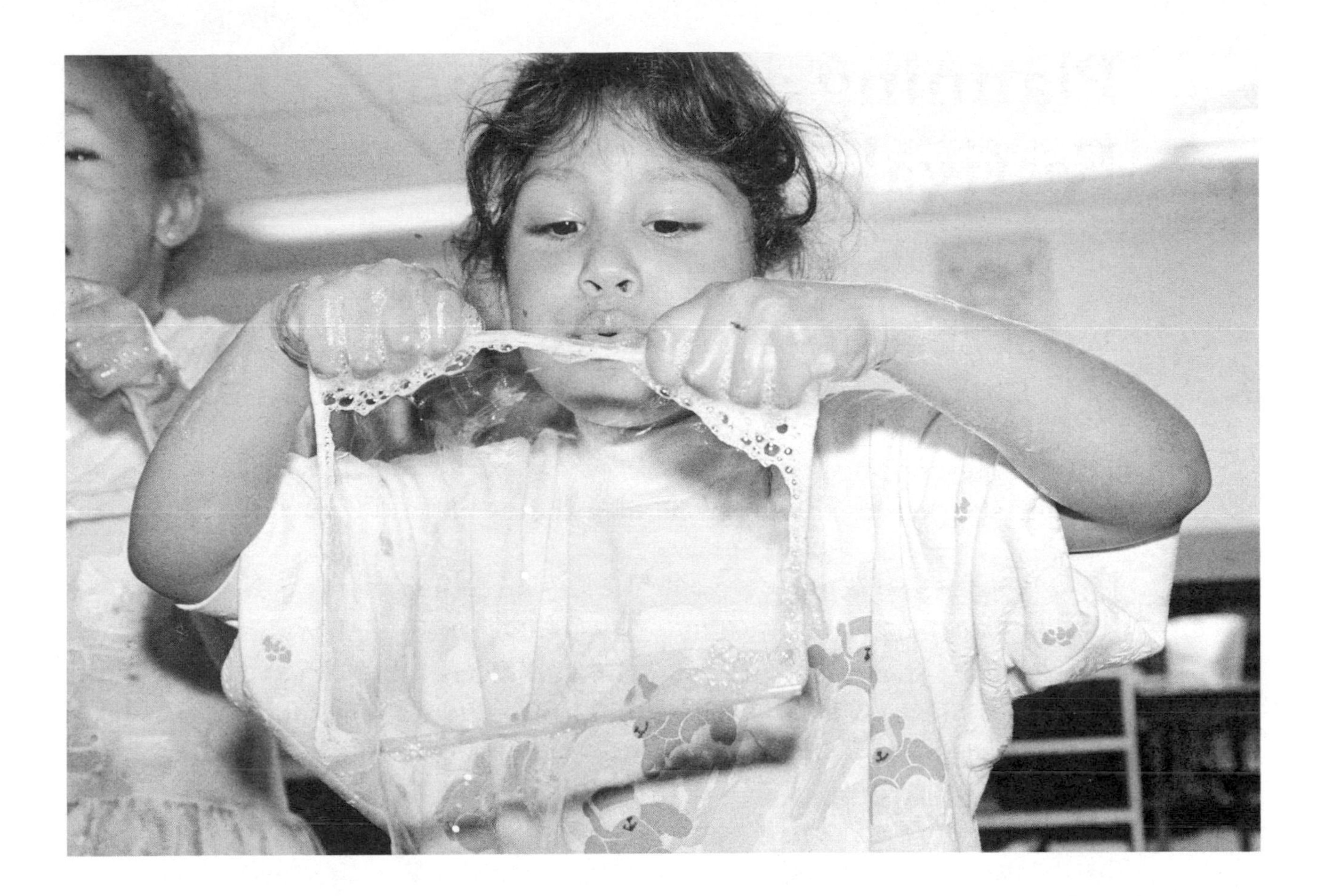

The First Session: Visit to the Pond

In the first session of a *Bubble Festival* your students learn to become successful bubble blowers, discover and explore what interests them about bubbles, and learn your expectations of them. Imagine taking a group of 30 students to a pond, and asking them to focus immediately on counting the facets of a dragonfly's eye. It's an impossible request! Children need time to run around the pond, poke a stick in the pond to estimate its depth, discover all the different pond life they can, and to skip a few rocks. On a subsequent visit to the pond, you can start to expect your students to settle down and focus on specific things. The same is true with a *Bubble Festival*.

For this reason, many teachers find that a successful way to launch a *Bubble Festival*, is to have six learning stations all with the same activity, so students have time to explore and experiment freely. The first two activities ("Body Bubbles" or "Bubble Shapes") are particularly well suited to being used in this way, as they are very open-ended by nature. The next two activities ("Bubble Measurement" or "Bubble Technology") also can be used successfully as a first encounter with bubbles, although they require a little more focus and more materials than the first two activities. Keep in mind that it is not uncommon during a first session for students to lose track of the challenge at a station. For instance, they will be very engaged in blowing bubbles on a table at the

"Bubble Measurement" station but even with lots of encouragement, they may not measure the bubbles. Activities 5-12 are not recommended as a sole, first encounter.

Some teachers prefer to set their students loose with straws, bubble solution, and tables, and let them explore with these materials for an entire session, before encountering *any* specific challenges. A few teachers have found that having six different learning stations during the first session, and rotating students through the stations after only 5-10 minutes, can also serve the purpose of allowing students to "run around the pond," and has the advantage of allowing the teacher to get a feel for which were the students' favorite activities and to use that information in planning the remaining sessions.

The format you choose for your first session is therefore best decided by you, with your knowledge of your particular students, and your preferred instructional style. However you choose to proceed, remember that you will need to drop all expectations for specific learning outcomes, and appreciate this first "visit to the pond" with all of the diverse and unanticipated learning that naturally occurs.

Subsequent Sessions

There are as many ways to format the subsequent sessions of a *Bubble Festival* as there are teachers! Some teachers choose the same format for each session, while other teachers use a different format each time. While the most popular format is to have two or three different activities per session, there are different ways to group activities within a particular session, and different ways to sequence them from session to session. Following are some important points to consider as you plan how you will organize your *Bubble Festival*.

1. The more activities you select per session, the greater the amount of set-up time will be required. While the first three activities ("Body Bubbles," "Bubble Shapes," and "Bubble Measurement") are extremely simple to set up, some of the other activities are a little more involved. Also, keep in mind that the more activities per session, the less time your students will have to explore at each station.

2. Certain activities are easier to duplicate than others. These activities ("Body Bubbles," "Bubble Shapes," and "Bubble Measurement") require very few and simple materials and can just as easily be set up at six stations as at one station. Other activities have more materials and materials that take some preparation, but can still be duplicated easily with some advance planning. Activities 9-12 are listed in the category of "More Elaborate Learning Stations" and if you decide to

include any of these wonderful activities in your *Bubble Festival*, you will probably want to limit them to one or two stations.

Some teachers find it easiest from a preparation perspective to set up the same selection of activities during each session, and organize groups of students so that on a given day they focus on two of the activities. By the end of three sessions they have had a chance to interact with all activities.

3

Some activities are significantly messier, requiring more clean-up time than others. Specifically, the airborne bubbles are messier than the table bubbles, and "Bubble Windows" wins the prize for being the messiest! The good news is that clean-up time does not change significantly as the number of activities per session increases.

4

Some activities are better presented by themselves, or with an activity that is equally attractive to students. Certain activities are more exciting to students than others, and tend to draw student interest away from the other stations. "Bubble Windows" is a favorite for all ages of students, and is best presented alone or with another super exciting activity, such as "Bubble Walls" or "Swimming Pool Bubbles." Likewise, "Bubble Colors," "Bubble Skeletons," "Stacking Bubbles," and "Frozen Bubbles" are more thoughtful, quiet, focused stations, and work well when by themselves or grouped with each other.

5

Consider the excitement level of each activity as you plan how to sequence activities from session to session. Some teachers feel more comfortable in going from calm to excitement, so the *Bubble Festival* builds to a crescendo. Other teachers prefer to start out with the most exciting activities in first sessions, to match their students' hard-to-contain excitement that naturally comes with early encounters with bubbles, and then to work up to the quieter, more thoughtful activities in later sessions, when students have become accustomed to being around soap bubbles, and more attentive to fine observation.

6

Consider featuring a certain process skill, theme, or other content-full way of grouping activities. Some teachers choose to feature a process skill such as observing in one session, predicting in another, and hypothesizing in a third. Two or three activities that lend themselves well to each of these process skills are featured on each of those days. Other teachers have begun with activities that have airborne bubbles, gone on to activities that use table bubbles, and finish with those activities that feature unusual bubbles. Another way to plan sessions is to choose activities on one day which focus students on ways that bubbles change and another day on ways they stay the same. A variety of specific ways to group activities appears in the next section.

7

Keep in mind that there are reasons you may want to repeat a certain activity. There may be a favorite activity that students can't get enough of. Or you may decide to bring out an activity a second time, when your students are more able to focus, or when you have focused them in a different way.

8

You may be able to have one of these activities set up in an ongoing, rotating fashion, in an area of the classroom where students have choice activities. Needless to say, it is difficult to have some students blowing bubbles while others are working on, say, spelling. We wouldn't have believed it at the start, but there are teachers who are able to use some of these activities in an extremely successful way as they do their other learning station activities. After an initial whole class session, which helps dissipate some of the extreme student excitement, these teachers are able to rotate a small group of students through one bubble station while other students do something else, or include a bubble station in a menu of learning station activities from which students can choose to spend time. The teachers who make this work have set up an environment in which students are accustomed to working in small groups at learning stations, and have an array of highly appealing stations from which to choose. If you decide to proceed with this format, do so cautiously, as bubbles are unbelievably absorbing.

Planning Your Festival

Clearly, there are a tremendous number of variables to consider in planning the format of your *Bubble Festival*, and a large number of successful models. Our goal is to communicate the collective experience of the many teachers who have presented *Bubble Festival*s so you can decide what's important to you and plan your festival accordingly. Whatever you decide, have fun with it, and don't be afraid to change your plan mid-course as you make your own discoveries. Here are a variety of formats that have been successful for teachers. These specific ideas can get you started in your planning.

- **Assign one month to be Bubble Month, and allow enough time to do preliminary activities, learning stations, and follow-up activities.**

- **Set it up and do it all in one week!**

- **Spread it out over a month or two, saving Thursday afternoons as *Bubble Festival* time.**

Here are six examples of *different* ways
you could plan and structure *Bubble Festival* class sessions,
but also feel free to come up with your own!

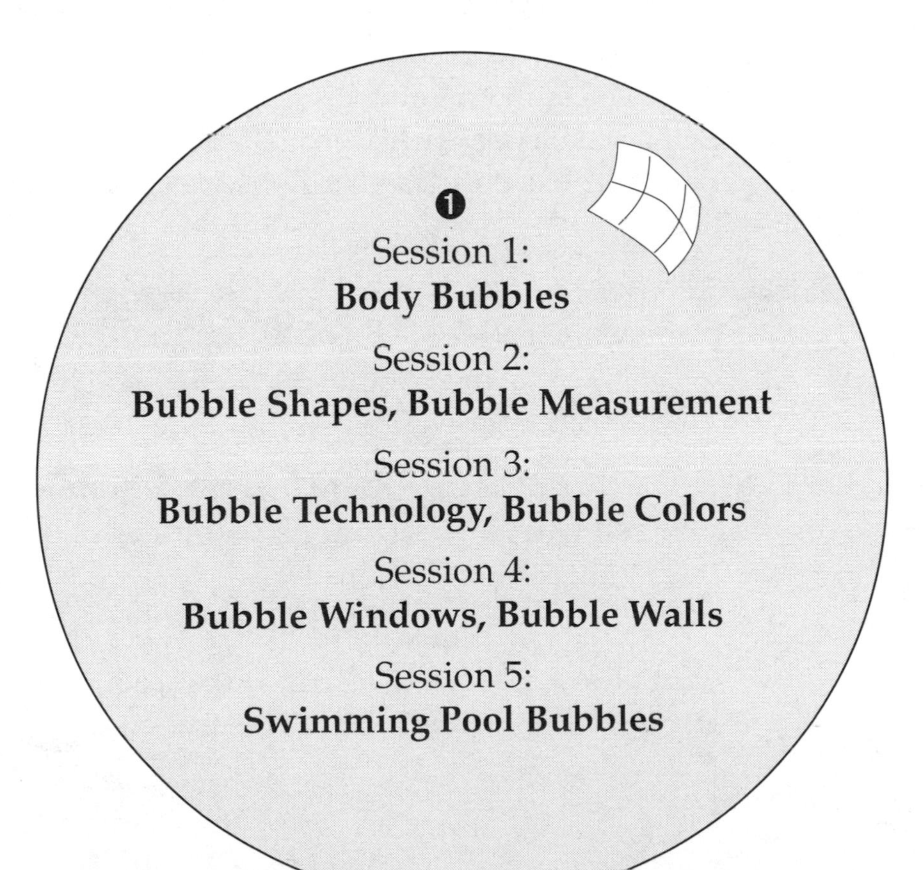

❷

Session 1:
Exploration (Bubble Measurement)

Session 2:
Structure of Bubbles (Bubble Shapes, Bubble Measurement, Bubble Skeletons, Stacking Bubbles)

Session 3:
Different Ways of Making Bubbles (Body Bubbles, Bubble Technology, Bubble Windows, Bubble Walls, Bubble Foam)

❸

Session 1:
Exploration (bubble solution and straws)

Session 2:
What's the Same About Bubbles? (Bubble Shapes, Bubble Measurement, Bubble Technology, Bubble Foam)

Session 3:
How Do Bubbles Change? (Bubble Colors, Frozen Bubbles, Bubble Skeletons, Bubble Measurement)

Session 4:
Fantastic Bubbles (Bubble Windows, Bubble Walls, Swimming Pool Bubbles)

❹

Session 1:
Exploration (Body Bubbles, Bubble Shapes, Bubble Technology)

Session 2:
Sheets of Soap Film (Bubble Windows, Bubble Walls)

Session 3:
Spherical Soap Film (Body Bubbles, Bubble Measurement, Bubble Colors, Bubble Shapes)

Session 4:
Cylindrical Soap Film (Swimming Pool Bubbles)

❺

Session 1:
Table Bubbles (Bubble Shapes, Bubble Measurement, Bubble Colors)

Session 2:
Airborne Bubbles (Body Bubbles, Bubble Technology)

Session 3:
Other Bubbles (Bubble Windows, Bubble Walls)

❻

Session 1:
Body Bubbles

Session 2:
Bubble Shapes

Session 3:
Bubble Measurement

Session 4:
Bubble Technology

Session 5:
All of the above stations with the class hosting a younger group of students at their *Bubble Festival*

Where to Have Your *Bubble Festival*

Outside? No!

By now you probably have thought about the potential mess created by 30 students working with bubble solution, and have decided to conduct your *Bubble Festival* outside. As appealing as this may sound, we do not recommend that you have your *Bubble Festival* outside. In our experience, 19 out of 20 times, an outdoor *Bubble Festival* will be unsuccessful.

- Bubbles are broken or blown away with the slightest breeze.

- Blazing sun causes the water in the soap solution to evaporate faster than usual, which causes bubbles to pop sooner than they should.

- Certain activities just won't work, and others will be reduced to an observation of the interaction of air currents and soap solution. This can be fun and educational as a one-time activity but will tend to remove the specific focus of almost all of the activities described in this guide.

- Even if you have an extremely still and humid day, preferably overcast, in a protected area, where tables can be moved and placed on a level surface and want to try the festival outside, be prepared for it not to work. We have set up *Bubble Festival*s in protected courtyards to find that an unnoticed wind tunnel existed, and then have had to move the whole thing back inside. Other times, conditions were good at the beginning of a *Bubble Festival*, and then the wind would pick up, and midway through a *Bubble Festival*, we had to move tables inside. In these instances, trying to have a *Bubble Festival* outside made for more work.

- Also, a *Bubble Festival* conducted in the open may attract every person in the vicinity, requiring you to spend your time keeping these people from joining, or trying to accommodate them at overcrowded stations.

Therefore, while the appeal of avoiding some of the clean-up by holding the festival outdoors is strong, keep in mind that conducting a *Bubble Festival* outside is fraught with potential problems.

Inside? Yes!

A cafeteria or multi-purpose room will work well if you can set up, do the activity, and clean up within a schedule that does not conflict with the needs of others at your school. While you will still need to tend to the table surfaces,

and drips and spills on the floor, you will not need to worry about books, fish tanks, bulletin board displays, or student papers that may become "anointed" with bubble solution. There is also the advantage, in large rooms, of students being able to spread out to different areas.

Many teachers find that their own classrooms are often the best locations for *Bubble Festivals*. If you conduct your *Bubble Festival* in your classroom, or another room containing things that might need protection from bubble solution, there are several important, simple precautions to take. These are described in detail on page 37 in "Preparing the Room and Materials."

Flat Surfaces

Wherever you locate your *Bubble Festival,* you will need to dedicate six flat surfaces at least 2 x 3 feet to serve as bubble stations for 4–6 students each. Tables are the best option, although you can also use counters or student desks. Slanted desktops can be leveled with books. Most table and desk surfaces are very easy to clean—in fact the soap solution will leave these surfaces sparkling clean! The important thing is that students have room to work comfortably at the station. For details on how to arrange and protect the contents of students' desks see page 40 in "Preparing the Room and Materials."

The Floor

There are two good reasons to be concerned about the floor during your *Bubble Festival*: safety and clean-up. A slippery hard floor is a serious hazard, especially with 30 excited kids in the room, and soap solution is harder than other liquids to remove from a floor. Both of these concerns are addressed completely on page 45 in "Tips for Clean-Up." There are ways to limit the amount of bubble solution that goes on the floor in the first place, and these are discussed on page 42 in "Tips on Organization and Management."

Carpeted Floor Surfaces

In selecting a location for your *Bubble Festival,* keep in mind that there are many advantages to conducting it on a carpet or rug. This may be hard to believe, but consider the following: A carpeted surface does not become slippery when wet, a modest amount of bubble solution spilled on carpet need not even be cleaned up as generally it dries fairly quickly, and leaves only a clean-smelling residue. Think of when rugs are shampooed—they are never rinsed of soap completely. The remaining soap residue is left in the carpet.

If a huge amount of soap solution is spilled on the rug, which will not dry, the excess solution can be blotted up by standing on layers of paper towels placed on top of the carpet. Most carpets, especially the kinds found in schools

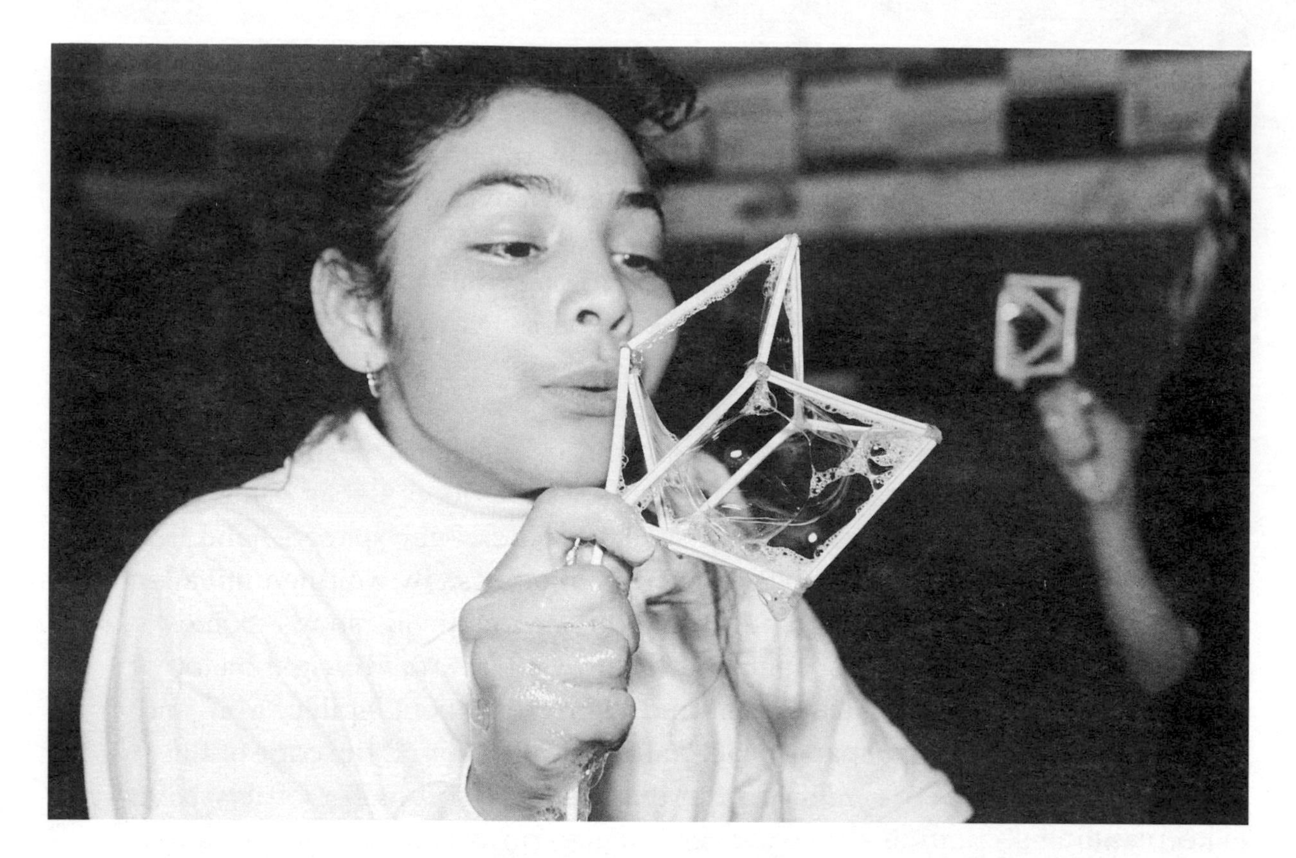

and community centers, are treated with mildew-resistant chemicals, so even if the carpet remains damp for several days, no mildew will form. Indoor/outdoor carpeting is an ideal floor surface on which to have a *Bubble Festival*. We have taken *Bubble Festival*s out to hundreds of schools, bringing strips of indoor/outdoor carpeting to protect and make bare floors safe.

Bare Floor Surfaces

Your goal, if you have a linoleum or wood floor surface in the room where you conduct your *Bubble Festival*, is mainly to keep it safe and to *reduce* the clean-up. Do not attempt to cover the floor of the room with a thin layer of newspaper. Not all of the floor will need protection. But more importantly, this will take you a long time to set out, absorb very little, and will quickly be trampled, torn, and crumpled by 30 pairs of feet!

Many teachers feel comfortable not protecting the floor at all until spills occur. This is definitely the easiest, least time consuming method and seems to work well for many people. If you are very concerned about your floor and safety, and have the time and help to cover floors prior to your *Bubble Festival*, then proceed as follows. First, identify those stations that are at highest risk of wetting the floor ("Bubble Windows," "Bubble Technology," "Bubble Walls," "Bubble Foam," "Bubble Skeletons," and "Swimming Pool Bubbles"). Then spread a drop cloth or old bed sheet beneath the edge of the table at which one of these stations is set up, and use some duct tape to anchor it.

Most likely, you won't have access to half a dozen drop cloths, so you may want to use butcher paper. While butcher paper isn't very absorbent, it keeps the floor from being slick when wet, and won't be tracked around like newspaper. A roll of butcher paper can be used to quickly cover the length of the front edge of a table. Use masking tape or duct tape to tape down the edges of the paper. Don't spend a long time making a perfect tape job—just tape enough to anchor the paper and keep feet from going under it. Again, covering the floors prior to a *Bubble Festival* is optional.

Whether you attempt to cover the floor prior to your *Bubble Festival* or not, you will need to be prepared to deal with spills. Bare floors + soap solution + fast moving children = a hazard! Have several stacks of newspapers handy, so you can throw a section or two on a spill when it occurs. By waiting until after a section of floor is wet, the newspaper you throw down will "stick." Some teachers carry a stack of newspaper over the arm, so they can be right on top of spills. Better yet, advise your students and volunteers about "spill patrol," and have several stacks of newspaper placed strategically around the edge of the room, so spills can be covered soon after they happen. You will probably have to keep adding sections as the *Bubble Festival* goes on, and removing crumpled newspapers that get in students' way.

As for removing soap solution from bare floors after a *Bubble Festival*, we have some advice on page 45 in "Tips for Clean-Up." It does not require a major effort.

While bubble solution can be messy, it is certainly not dirty! If care is taken to remove or protect certain items prior to a *Bubble Festival*, the issue is solely one of the time it takes to clean up. Clean-up issues are addressed on page 45 in "Tips for Clean-Up." Once the clean-up is complete, you, your students, and their desks will be far cleaner than when you started!

Using the Signs

In the "Getting Ready" section on page 37, we describe a simple way of making bubble proof, stand-alone signs, which guide your students in their investigations at each station. Please note that for two stations, "Bubble Measurement" and "Bubble Shapes" we have provided alternate signs for younger students. In general, the signs have a minimum of words. Be advised, however, that your students will still not want to stop and read the signs during the *Bubble Festival*. Blowing bubbles is far more interesting than reading about them. Teachers solve this problem in a variety of ways.

GEMS is in the process of translating the Bubble Festival signs into Spanish; they should be available in late 1994.

✌ Some teachers choose to read the signs aloud to students, at the beginning of each session, before the students begin blowing bubbles. This method is especially effective when there are no more than three activities per session. It works well with non-readers, who then have the pictures to remind them later.

✌ One teacher read the signs aloud to the students on the day prior to the activity, and recorded her students' predictions about what would occur. When the students embarked on the activities the next day, they had good recall, as they were testing their predictions.

✌ Another teacher who uses cooperative groupings, had each group meet for 3-4 minutes to read over the signs and discuss their job as a team. This worked very well to focus them and help them understand that they had a job to do.

✌ Hand coloring the signs can make them more interesting to students and the color can be used to highlight the most important aspect of a sign.

✌ Other teachers find that by constantly encouraging and reminding students to read the signs, they have a fairly high degree of success. Some teachers set up a procedure for students to follow, such as the following:

1) Go to your station,
2) Read the sign aloud in unison with a partner,
3) Do what the sign says. Then remind their students about "1, 2,3."

✌ You may want to motivate your students to read the signs by letting them know of a sharing time that will follow the *Bubble Festival* at which you would like them to report their findings.

✌ Some teachers use the signs purely to inform adult volunteers of the goals of a station. Then the adults can help guide the students' investigations by asking questions, and demonstrating a technique if necessary.

✌ You may have another system for guiding students at each station and may decide simply not to use the signs.

Getting Help

During the *Bubble Festival*

Most teachers find that a *Bubble Festival* is far more successful and enjoyable if they can enlist the assistance of other adults. Having just one parent volunteer to focus on monitoring the amount of soap solution that accumulates on the tables (thereby greatly reducing drips and spills), and keeping the floor safe, will enable you to concentrate on the students, their discoveries, and their behavior. Having six parent volunteers (one at each station) makes a *Bubble Festival* run *very* smoothly, and is especially desirable with younger students. In general, there is agreement that the more adults present, the better. Some teachers have set up two tables of the same activity next to each other, so that one adult can easily monitor two tables. If most of the parents in your school community work, then consider inviting grandparents, or contacting a local branch of the American Association of Retired People. It has been our experience that adult volunteers enjoy the experience of *Bubble Festival* nearly as much as the children!

To help you get the best help from adult volunteers, we have included "Activity Task Cards for Volunteers" after each Learning Station Activity. These briefly explain the learning goals of each station, suggest additional questions to ask of the students, and outline how to maintain the station so it is safe and ready for the next group of students.

While for the most part, the adult help you'll need requires no special preparation or skill, you may find a parent or two who interferes with students' open-ended investigations by demonstrating and showing students too much. Some teachers have found that a short orientation with adult volunteers is time well spent. This can be as brief as a few sentences to the adults as they come in, or when you speak with them on the phone. Even if you haven't had time to mention anything to the volunteers before the *Bubble Festival*, you can mention something as you introduce the *Bubble Festival* to your students. You might tell the students when the parents are listening, "We adults will help remind you of the goal at each station, but *you'll* be the scientists today. You'll be the ones to investigate the questions at each station and make your own discoveries."

While extra adult assistance during a *Bubble Festival* is highly desirable, many teachers (especially of older students) have managed to conduct an entire series of *Bubble Festival*s with their students quite successfully without extra help.

Several teachers have paired older classes with younger classes to the delight of all involved. This works especially well if the older students have had a chance to experience a session or two of *Bubble Festival* by themselves, first. One class of 4th graders even hosted a mini-*Bubble Festival* for the kindergarteners at their school. They planned the set-up and the clean-up as a group, and talked about the best ways to help the kindergarteners enjoy and learn.

Before and After the *Bubble Festival*

If you plan correctly, you can also enlist the parent volunteers' help in setting up and cleaning up the *Bubble Festival*. As we will describe later, while students can be involved in some aspects of the clean-up, at a certain point it is ideal to be able to remove most of the students from the room, and have several people finish up. The easiest way to achieve this ideal situation is to arrange to take your students to the playground, to the library to read a story, or some other logical place, while the parent volunteer(s) complete the clean-up.

Many teachers found that by working together to present *Bubble Festivals* to two or more classes, the set-up and clean-up can be significantly reduced. A successful model for this is to set up the *Bubble Festival* in one room, and then rotate different classes through the room over the course of the morning. In this way, there is only one set-up and clean-up required per day, and two or more teachers available to do it. See "Presenting *Bubble Festivals* to Several Hundred People" on page 125 for details of how one school cycled six classrooms through a multi-disciplinary *Bubble Festival*. A *Bubble Festival* is an ideal collaborative project.

Grade Level Considerations

Our experience has shown that *Bubble Festival* is a hit with kids from 3 to 93. While of course, there are modifications which make the activities more successful with students at a specific grade-level, everybody (including adults!) needs a chance to explore freely in the beginning. Younger students may need more free exploration than older students.

After your students are ready to move beyond free exploration, they can be asked to make and test predictions, to record the results of their experiments, and to write about their experiences—this is especially appropriate for older students. Some 5th and 6th grade teachers have had their students write and present their scientific results to the rest of the class, or "publish" them as part of a Bubble Science Journal.

Younger students can certainly share and record their experiences as well, but keep in mind that initially, just the challenge of making a bubble is great when you're only 5 or 6 years old! Some teachers of Kindergarten and 1st grade choose to have several larger tubes on hand, in case some of their students find the straws too frustrating to use. Alternatively, having more parent volunteers can greatly aid the youngest students in being successful sooner.

In general, teachers of younger students need to be prepared for more soap in eyes and mouths. Remember to consider the height of your students as you choose the location of such activities as Bubble Walls (students need to be able to lift the dowel to its full height), or Bubble Colors (students need to see over the bubble homes).

The activities themselves are well-suited to a range of ages. Students naturally interact with the materials as appropriate to their experience and background. In two cases, we have included a more elementary sign to be used with younger students. Feel free to modify signs and activities according to the experience and interest of you and your students.

What You Need

For About 3 Sessions of a *Bubble Festival*

** Items with an asterisk are consumable and will need to be increased for more than approximately three sessions with one class of about 30 students. All other items on this list can be reused.*

One parent volunteer wore her plastic raincoat! This is going a bit far, but she did remain dry.

✓ lots of water*

✓ 2 one-quart containers of Dawn® or Joy® dish-washing liquid*

✓ 2 cups glycerin (Glycerin is available at well-stocked pharmacies and scientific supply companies.)*

✓ 1 or 2 packages of paper towels*

One teacher brought in some old large bath towels to use during her Bubble Festival. She used them on the floor and to wipe hands. She sent them home with parents to be laundered. When Lawrence Hall of Science takes these activities to schools, we provide our Bubble Festival volunteers with cotton tea towels. Stuffed in a back pocket or hung over a belt, they can be incredibly useful, as wipers or blotters.

✓150 drinking straws (Small diameter straws don't work well. Some teachers like non-clear straws as they are easier to keep track of. Clear straws enable you to actually see if a student who is having difficulty blowing bubbles is inhaling first, and thus breaking the soap film at the end of the straw.)*

✓ 2-4 five-gallon buckets for mixing bubble solution

✓ 1 empty one-gallon container

✓ 1 one- or two-cup capacity measure

✓ 3 or more squeegees (While you can get by with one squeegee, having multiple squeegees will enable you to have your students and/or parent volunteers quickly remove the soap solution from station surfaces, both during the *Bubble Festival* to reduce spills, and after the *Bubble Festival* to facilitate clean-up.)

✓ 1 plastic squirt bottle (an empty dish-washing liquid bottle works well)

✓ 1 quart vinegar*

✓ 3 dish pans for clean-up

✓ drop cloths, butcher paper*, and/or a stack of newspaper* to absorb spills

✓access to a laminator (Laminators are sometimes available at copy centers.) Lots of clear contact paper can be used to protect the signs from bubble solution, but a small amount of leakage will occur. A better method using two file folders, glue, and rubber cement, followed by lamination, is illustrated on page 38.

Optional

1 small rubber spatula

For the Individual Stations

A list of the materials you need for each activity is found on the page for that activity under "What You Need for One Station." If you are setting up more than one learning station for an activity, just multiply the materials listed by the number of stations you need.

Getting Ready

Preparing the Room and Materials

Before the Day of the Festival

☞ Acquire all of the necessary materials.

☞ Mix Bubble Solution. You will need approximately 16 gallons of bubble solution to conduct three sessions with one class of students. (This is very approximate, as some stations use up more solution than others, and some students use up more solution than others!) This can be easily made in four 5-gallon buckets. If you have just two 5-gallon buckets, then start by making two buckets of bubble solution before the day of the activity, then make more as you need it.

Bubble solution keeps well. In fact, it seems to improve significantly with age! Teachers who have made their solution several weeks in advance have commented that the "aged solution" seems to make bigger, sturdier, more elastic bubbles. If you do keep your solution for more than a day or two, put a lid or cover on the bucket to help prevent evaporation.

Start by filling each 5-gallon bucket with 4 gallons of water. After adding the water, measure and add 4 cups of dish-washing liquid to each bucket. Measure and add one-half cup glycerin to each bucket. (Use a small rubber spatula to wipe the glycerin out of the cup.) Stir the solutions gently with your hand or with a yardstick that you don't mind getting wet.

RECIPE FOR BUBBLE SOLUTION

Mix the bubble solution near the place it will be used if possible.
Remember that water weighs 8 lbs. per gallon!

4 gallons water
4 cups Dawn® or Joy® dish-washing liquid
½ cup glycerin

☞ Make Signs. Xerox the two-page master signs following each activity you have selected and laminate them onto file folders. Make sure the seal is good, as sitting in a puddle of bubble solution is the ultimate test of this waterproofing technique! You will need one sign for each station for each activity. If you have chosen to read signs aloud to students and not have the actual signs at the stations, then there is no need to laminate them. Some teachers, who wanted to skip the lamination step, but use the signs at the stations, found places to put the signs where they would not come into contact with bubble solution. (One teacher hung the signs, another stuck them with magnets to the chalkboard, still another teacher had wire frames on which he set the signs.) If the signs will come into contact with bubble solution, you will save a lot of time in the long run by making the effort to laminate the signs now. They can be easily stored in your *Bubble Festival* files.

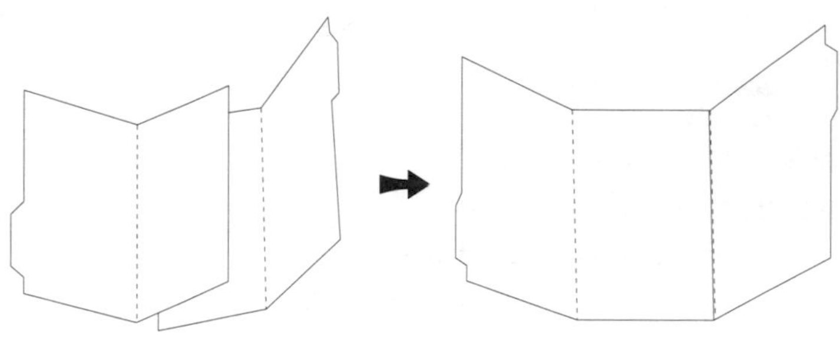

① Glue two folders together so that you have a three-panel sign backing. (Use any paper glue.)

② Use rubber cement (best because the pages won't "bubble up" off the file folders) to attach xeroxed signs to backing. Optional: Color the signs.

③ Laminate signs.

④ Label a tab or edge of the sign for easy retrieval from your *Bubble Festival* file.

If you do not have access to a laminator, you can use clear contact paper to cover both sides of the sign. If you do this, make sure to lap the edges of the contact paper as pictured, so the bubble solution will not leak through. Even so, some leakage is likely and lamination is the recommended method.

① Cover the back of the sign with contact paper.

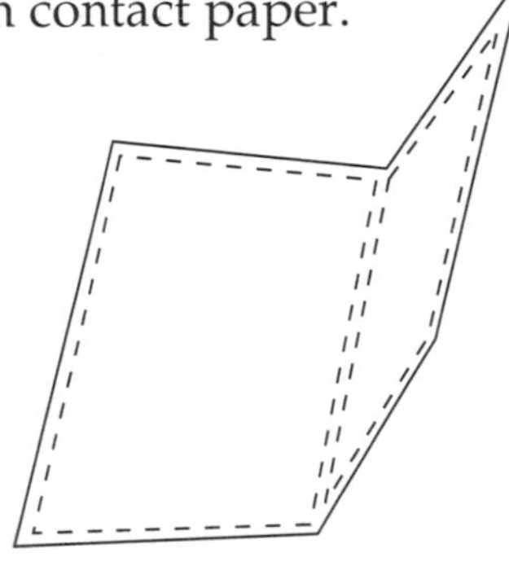

② Cover the front of the sign so that the contact paper seam is not on the bottom edge of the sign. Make the contact paper a little longer than the sign and fold it up around the bottom edge. This helps prevent leakage.

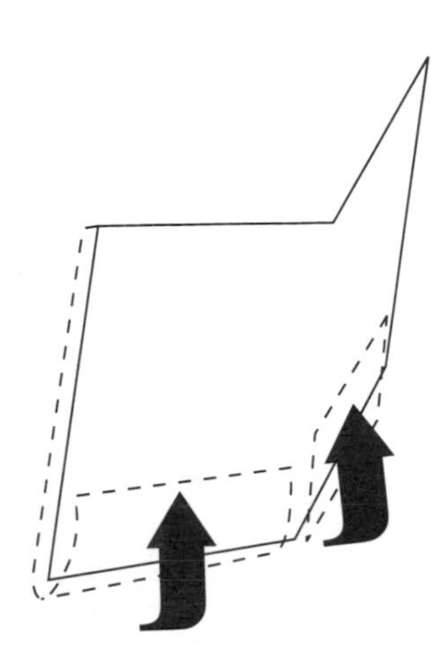

☞ Xerox task cards. If you are using parent volunteers, you may want to duplicate the task cards near the end of this guide, and cut them up for easy distribution to volunteers, or, if possible, laminate them.

On the Day of the Festival

Many teachers have found that presenting a *Bubble Festival* first thing in the day, after lunch, or after a preparation period, enables them to set up the room and the materials in a calmer setting. If possible, schedule it in this way.

☞ Prepare the room and plan where your stations will be.

- Before a *Bubble Festival* cover the top of a classroom fish tank, or that of any aquatic animal living in your classroom, with butcher paper or a cloth.

- Make sure to place bubble stations away from bulletin boards or other wall displays to protect them from bubble solution. If you must place a station near one of these areas, make sure that it involves table bubbles or one of the more contained activities.

- "Bubble Windows," "Bubble Technology," "Bubble Walls," "Bubble Foam," "Bubble Skeletons," and "Swimming Pool Bubbles" are the activities with the biggest potential for mess. If possible, locate these near a sink, or an area that is easiest to clean.

- Some teachers prefer to have their students stack their chairs and place them out of the way. Not only is it best if students stand at the stations, but having the chairs out of the way will make one less thing to wipe off later.

Another thing to consider as you arrange the room is that "Bubble Windows," "Bubble Walls" and "Swimming Pool Bubbles" are particularly sensitive to drafts, so choose stations that are located as far as possible from drafts, such as doors, windows, or air vents.

☞ Arrange the tables or desks so there are six flat surfaces at which to locate stations. If you are using student desks, they must be level. They can be pushed together to make larger work areas. Some desks can be turned inward, with the open side facing the open side of another desk, so bubble solution will not dribble into its contents.

Clear each station surface, removing anything that would be damaged by soap solution. If a station surface needs protection because there are stickers or other paper items adhered or contact-papered to the surface, then cover it with a plastic trash bag, as described in "Bubble Colors" on page 80.

☞ If there is no carpet, cover those areas of the floor that are most likely to become slippery. First, identify those stations that are at highest risk of wetting the floor: "Bubble Windows," "Bubble Technology," "Bubble Walls," "Bubble Foam," "Bubble Skeletons," and "Swimming Pool Bubbles." These are the only stations that need floor protection. If your room has carpeting, floor covering is unnecessary.

If possible, spread a drop cloth or old bed sheet beneath the edge of the table, at which one of these stations is located and use some duct tape to anchor it. If you don't have drop cloths, a roll of butcher paper can be used to quickly cover the length of the front edge of a table. Use masking tape or duct tape to tape down the edges of the paper. Don't spend a long time making a perfect tape job—just tape enough to anchor the paper and keep feet from going under it. Newspaper can also be used, but it should be put down at the appropriate places in stacks (at least a section thick). It is difficult to anchor the newspaper well, so you will probably have to keep adding sections as the *Bubble Festival* goes on, and removing crumpled newspapers that get in students' way.

☞ Set up learning stations. Put out the equipment, bubble solution, and signs for each learning station. Have stacks of newspapers handy near the stations.

On the table there should be two cottage cheese containers about half full of bubble solution, a measuring tape, a ruler, a protractor, some uniform small cubes (about an inch by an inch), a few popsicle sticks, a few toothpicks, and two pieces of string—each about a yard long.

☞ Prepare and set aside your clean-up materials. In the plastic squirt bottle, mix a solution of half vinegar and half water. Place the squirt bottle, squeegees, three dish pans, and the paper towels out of sight in a place where you can find them when it is time to clean up.

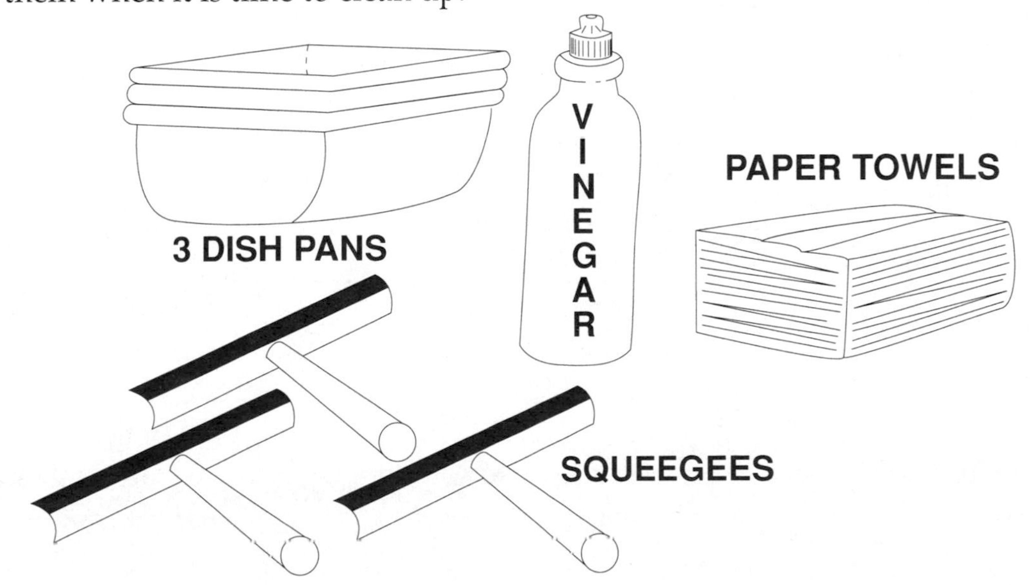

☞ Prepare and set aside your demonstration materials. For your demonstration, have a container of bubble solution, a straw, and a paper towel handy. The demonstration will need to take place in the area where you will gather the students just before the *Bubble Festival*. Put the demonstration materials on a table, cart or other flat, waterproof surface that students will be able to see. Practice the demonstrations described below on page 47 in "Introducing the *Bubble Festival* to Your Class" ahead of time.

Tips on Organization and Management

Getting Your Students' Attention

Before the day of the *Bubble Festival*, give some thought to a *Bubble Festival* management system that suits your usual style. It is *essential* to have a way to get the students' attention, whether it's flicking the lights on and off, ringing a bell or saying a certain phrase. This may seem obvious, but getting your students' attention while they're engaged in blowing bubbles can be very difficult.

Moving From Station to Station

Depending on the age of your students and on your individual class, plan how you can best direct the students and avoid crowding at the stations. Some teachers may choose to assign groups of 4 to 6 students to each station for the start of the *Bubble Festival*. These small groups can be made randomly by having the students "count off." Alternatively, a teacher may decide to plan the groups carefully to balance students' learning styles or abilities. Student groups can be formed in whatever way you see fit.

If you feel that a system is needed to control traffic and ensure that no one station becomes too crowded, you may want to try a timed rotation during your *Bubble Festival*. With this system, students must stay at a station until you signal it's time to change. Depending on how many activities are available, you may also want to set up a sequence of stations that they visit in turn. Make sure your students (and you) are clear which groups will move to which stations. Because students will be involved in exciting investigations, your signal to change stations should always be preceded by a one-minute warning such as, "In one minute, everyone will be moving to a new station. Finish the bubble activity you're doing now so you'll be ready."

Even if you choose to have a scheduled rotation, try to have some time in at least one of the sessions of your *Bubble Festival* to let your students choose what activity they do. By letting students seek and choose their activities, we allow students to select learning tasks that are appropriate and interesting to them. A student who has chosen a particular task is naturally more committed and positive about it. We are all more likely to be motivated and enthusiastic about what we ourselves choose to do. We are more likely to comprehend and retain successfully what we have chosen to learn.

Limiting the Use of Straws and Paper Towels

To limit the number of straws used, and to avoid having them left all over the room, some teachers caution students to keep track of their straws. The teacher might keep extra straws in a pocket during the *Bubble Festival*, and discourage repeated requests for new ones. Similarly, to avoid wasting paper towels, some teachers don't make paper towels available until the end of the session, and they ask students to use only one.

Tips from Teachers

On the following pages is a collection of very specific tips from teachers for ways to make the preparation, organization, and management of kids and bubbles happen as smoothly as possible. Of course, setting up a situation that runs perfectly takes much more preparation time than some teachers may want to spend. So pick from the tips as they suit you.

- Number the stations consecutively so children move easily from one to the next. Hang large numbers from the ceiling so children know which station is which.

- Provide enough room for children to wave their arms at certain stations, such as "Body Bubbles," "Bubble Technology," "Bubble Windows," and "Bubble Skeletons."

- Mark straws with a dark pen or crayon about 1 inch from the end that goes into the solution.

- Have students clean up their stations, squeegee off excess solution, etc. just before they rotate. This limits spills and drips dramatically.

- Set up stations so that kids have to rotate one table to the left instead of across the room.

- Review safety and bubble basics at the beginning of each session.

- Set up stations on six trays or in tubs the day before the *Bubble Festival*. Then just bring out the trays as you introduce each station the next day.

✌ Designate a dry area in the room, and introduce the session here, and re-assemble the students there at the end of the session.

✌ Remind students that "bubbles hate dry things."

✌ Stop every 10 minutes and ask two students for a discovery they've made. This encourages sharing and puts the brakes on uncontrolled behavior.

✌ Foam is the "worst enemy" of a bubble! Alert students and parent volunteers to periodically remove suds from the surface of bubble solution in dish pans and cottage cheese containers. Caution students to dip rather than stir, as stirring and blowing in bubble solution containers makes foam.

✌ Use duct tape to anchor dish pans full of bubble solution to table tops. This will eliminate accidental spills.

✌ Send a note home, asking parents to send kids to school in "bubble friendly" clothes: short sleeves, able to get wet, with big pockets to hold straws.

Tips for Clean-up

1 TRASH
Discard used straws and paper towels.

2 FLOOR COVERINGS
Throw away wet newspaper and butcher paper from the floor. Gather up drop cloths if you used them, and either put them in a clothes dryer, or hang them in the sun to dry.

3 EQUIPMENT
Collect all equipment from the various stations. Rinse and set it out to dry; however, if it is not too soapy and/or if you will be using it again soon, it needn't be rinsed. Put a layer of paper towels on a counter, and set the soapy equipment here until you use it again. After repeated use (5 or 6 times), you will probably want to make sure that everything gets a good rinse.

4 TABLES
Remove bubble solution from the station surfaces by sliding it into a dish pan with a squeegee. Discard this soap solution as it is likely to be dirty from the tables. Squirt some vinegar solution (see page 41) on the table to cut the soap scum. Use one paper towel to spread the vinegar on all of the soapy surfaces, then squeegee the remaining vinegar solution into a dish pan and discard. You'll be amazed out how sparkling clean your tables are! *Note: In general, sponges are a detriment to cleaning tables. They just spread the soap solution around, and are difficult to rinse.*

5 FLOORS
On bare floors, use paper towels or rags to wipe bubble solution from the floor. On areas of the floor where there was a significant amount of bubble solution, you may want to spot mop with vinegar solution to cut the soap scum, however we recommend consulting with your custodian first, as he may not want you to use vinegar on certain floor finishes. On carpeted floors, areas with large spills can be cleaned by blotting up excess solution. Stand on layers of paper towels placed on top of the carpet. Discard the towels, and repeat until towels are only slightly damp.

While cleaning up a *Bubble Festival* seems fairly straightforward, there are several challenges which may not be apparent: getting the students to stop blowing bubbles, dealing with a certain level of chaos that naturally ensues as

students actively embark on a vigorous clean-up, and that there is an awful lot to clean up! Following are several general suggestions to help you meet these challenges:

- Have your *Bubble Festival* end at recess or another break, allow the majority of the class to leave, and keep a team of especially helpful students and/or parent volunteers to help you clean up.

- If you have parent volunteers helping you, you can take your students to the playground, library, or multi-purpose room for closure or to do some other activity while the parents remain in the room to clean up. If you have a garbage can near the door, students can throw away their straws and paper towels as they leave.

- If your class does stay in the room, you might want to avoid confusion and traffic problems by asking all but a few students to stay in their seats during clean up.

- If your tolerance for noise and activity is high, and your students are accustomed to clean-up activities, you might want to involve all of them in at least the first part of the clean up. Clearly define roles, so each student or group of students is in charge of a certain task.

- Don't be an over-zealous cleaner. Evaporation is on your side! Wet things will dry, so there's no need to hand dry them. If you are using the equipment again in the near future, there's no need to rinse things well, or in some cases to even rinse them at all—they will only get soapy again.

Introducing the *Bubble Festival* to Your Class

Before Students Go to the Learning Stations

✓ Gather the class away from the bubble stations. Have your demonstration materials handy. Explain that, before they start the activities, you need to give them some tips for success and safety. Say that your introduction may take 10 minutes, but it is very important, so they need to be patient. Tell them you'll keep your introduction as short as possible.

✓ Hold up a straw, and ask what it is. After they've told you, say that today it's not a straw, but a bubble blower. The difference, of course, is that you suck on a straw, but you blow on a bubble blower. Tell the class that if they do accidentally get a little bubble solution in their mouths, it will taste awful, but won't hurt them. If you have a sink, tell them they can rinse out their mouths to get rid of the soapy taste.

✓ Point to your ear and ask what it is. Explain that today, these flaps on the sides of our heads are also known as "bubble-blower holders." Say that each student will get a blower, and should keep it behind their "holder" when they aren't using it. Students who would rather keep their straws in their pockets may do that. In this way, straws won't be wasted.

Note: for students who are young enough to consider putting a straw ***in*** *their ear, do* ***not*** *suggest they use their ears as "bubble-blower holders."*

✓ Mention that you'll save paper towels by having everyone wait until the end of the *Bubble Festival* to dry their hands. Say that there are several reasons for this. First, most students just get their hands wet again, and get into a pattern of wetting and drying their hands throughout the session. Second, students who dry their hands might be reluctant to get them wet again, and they might miss some interesting fun.

✓ Explain that another reason why they should keep their hands wet during the festival, is that **BUBBLES HATE DRY THINGS!** Demonstrate by dipping your straw in bubble solution, and attempting to blow a bubble on your dry palm. (It won't work.) Then wet your palm with solution, dip the straw and try again. (It works.) Pop the bubble with a dry finger to show again how much bubbles hate dry things. Then blow a new palm bubble, wet a finger and poke it in the bubble. (The bubble won't pop!)

One teacher allowed her students to discover for themselves that bubbles hate dry things. This was an exciting discovery for her students and became one of the outcomes of the first exploration session. If you think your students are old enough to discover successful ways to blow bubbles without becoming too frustrated, you may want to add this nice twist.

✓ Explain that at some stations, students will be asked to make a "table bubble." Demonstrate how to make a table bubble:

1) Pour some soap solution on the surface of the table, and use your hand to wet an area about the size of a large pizza. (About 18" or 45 cm in diameter)

2) Dip a straw into the container of solution.

3) With the straw just touching the soapy surface of the table, gently blow through the straw to form a bubble dome, and continue blowing until it is the size you want or until it pops. Take more than one breath if necessary.

✓ As briefly as possible, explain to your students the system you have decided upon for managing the *Bubble Festival*. (Refer to "Tips for Organization and Management," on page 20.) If you are assigning groups to stations or having a timed rotation, explain that now.

✓ Finish by saying that there are five important rules for a safe, happy *Bubble Festival*:

1) Do not run. Don't even walk fast.

2) If you get soap solution in your eyes, it will sting for a while, but it is not dangerous. Blink your eyes, but don't rub them, because your hands are probably soapy. (If a student continues to be bothered, they could go to the sink and wash their hands, then splash water in their eyes.)

3) If you like to pop bubbles, pop your own!

4) Be cooperative and easy to get along with! Find ways to share materials. There is no whining or arguing necessary at a *Bubble Festival.*

5) Be **BUBBLE-OLOGISTS**! Ask yourself questions, and then experiment. Try something new.

✓ Explain the consequence of breaking a rule—whatever you've decided that will be. Remind students always to "freeze" and listen quietly whenever you give the signal. Ask if the students have any questions.

Note: These rules could be presented to the students on a day prior to the Bubble Festival. Then students can just be reminded of the rules at the end of this introduction. That will give students two chances to hear the rules, and will shorten the introduction. Some teachers write these rules on butcher paper and keep them posted throughout all sessions of the Bubble Festival.

✓ Distribute the bubble blowers (straws) and have the students begin.

Once Students Are at the Learning Stations

✓ Circulate during the session, ask focusing questions and offer to help students who are having trouble blowing bubbles. Here are three hints for successful bubble blowing:

1) Dip the blower before each attempt. Make sure the surface and every thing that touches the bubble is wet. You may want to suggest that they wet the entire length of the straw with their hands.

2) Blow *very* gently. It's hard to explain exactly how gently in descriptive terms. For a student who is having trouble, you might suggest that he blow "as soft as a mouse whisper." If that student is still having difficulty, you might have to blow a palm bubble in his soapy palm, so he can feel the correct amount of force the air going into the bubble should have.

3) Take a breath *before* you put your mouth on the blower. If you inhale with the bubble blower in your mouth, it may break the soap film at the end of the blower.

✓ Occasionally, you may find that a younger student will repeatedly suck on the straw, and get solution in her mouth. If you have such a student, ask for her bubble blower, saying that you can fix it so that it will still be a good bubble *blower*, but it won't work for sucking *in* the bubble solution. With a scissors, cut a small, diamond-shaped hole by making two snips about an inch from one end of the straw.

The extra hole will not interfere with bubble-blowing, but when the student sucks on it, only air will reach her mouth.

Note: Some kindergarten teachers choose to cut diamonds in all of the straws to avoid the problem entirely.

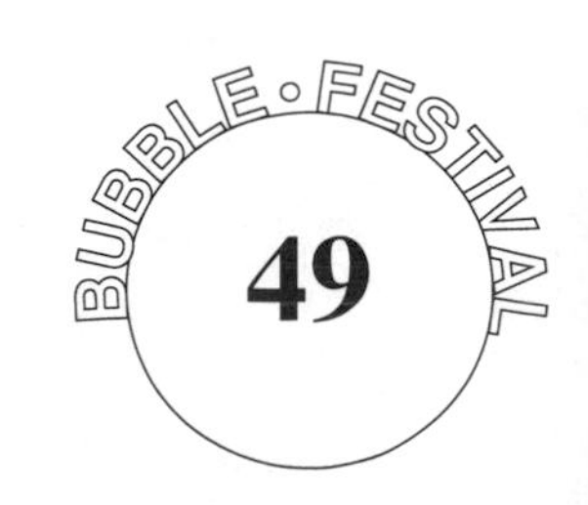

✓ Replenish the bubble solution in the cottage cheese containers and dish pans as needed. If there is a spill, help students absorb it with newspaper or towels.

Closure: Sharing Discoveries

There are many ways to provide closure to the experience of a *Bubble Festival*. Some teachers provide some kind of summation following each session. If it is difficult to conduct an orderly discussion, you could postpone the discussion to later in the day, or even first thing the next day. Other teachers wait until the end of all sessions to bring ideas together. Still other teachers have groups of students keep a bubble notebook in a dry area of the room, and jot down discoveries and questions during the session. Every 10 minutes a bell is rung and one student from a group shares one question or a discovery that had been made by their group. These questions and discoveries are then used as the basis for follow-up discussions and further experiments. Holding these discussions mid-way through the session also allows students to go back and try things that other students discovered. It lends a cooperative and collaborative sense to the Festival. Later on, all questions, observations, discoveries, and comments can be compiled in a class book.

Following are a list of different ways that teachers have successfully provided closure to *Bubble Festival* activities:

- Ask students what they discovered.
- Ask what new questions they have.
- Ask what they'd like to try again.
- Make a graph of the students' favorite activities.
- Ask for words to describe bubbles.
- Have your students draw and write about their experience with bubbles.
- Write several clue words on the board, such as "measurement," "colors," and "windows," and ask students to brainstorm words and phrases to list under each heading.
- Ask activity-specific questions that allow students to share their findings at each station. For instance, you might want to ask students what shapes they saw at the "Bubble Shapes" station and to draw them. Or you could ask students to tell you how they used their hands to make bubbles at the "Body Bubbles" station.
- Have your students write a letter to a friend or cartoon character, explaining how to blow a bubble. It is fun to actually test these step-by-step procedures the next time you have bubble solution out!
- Consult page 135 for a feature on "Writing and Bubbles" that includes lots of ideas for ways to incorporate writing into your *Bubble Festival*.
- Read a book which takes off on the students' experiences. (See page 136 for a listing of children's literature that relates to *Bubble Festival*.)
- Follow up festival sessions with some of the many "Going Further" activities suggested with each activity.

Learning Station Activities

with Activity Task Cards for Volunteers

Additional, removable copies of the Activity Station Signs and Volunteer Task Cards are provided at the end of this guide for your copying convenience.

Activity 1
Body Bubbles

Overview

Make bubbles with our bare hands? Yes! Some of the best bubbles are made with fingers, hands, and arms too, as long as they're wet with bubble solution. Make a circle with thumb and forefinger (an "okay" sign) or with the fingers of both hands, dip it in bubble solution, and blow bubbles on the table or in the air. Use a straw to blow bubbles in your palm. Cooperate with another bubble-ologist to find ways to combine your bubbles on the table or in the air. Find out how many people can blow into the same bubble.

One student found a way to blow bubbles all the way down his arm. He collected bubble solution in his hand and closed it. Then he put his straw into his hand and began blowing. As he blew, he closed his hand tighter and tighter. The bubbles began spewing out of his hand and down his arm. It looked like a bubble machine. Everyone enjoyed doing it.

What You Need for One Station

✓ 1 or 2 cottage cheese containers

Getting Ready

☞ Fill the cottage cheese containers about half-full of bubble solution.

☞ Set out the "Body Bubbles" sign.

Special Considerations

- This activity works best when students have a large surface on which to work.

- Squeegee off the table periodically.

Going Further

- Have your students write instructions for making the perfect body bubble.

- Give the homework assignment of taking a bubble bath. Ask each of your students to share one new discovery they made in the tub!

- Begin your students' writing with one of the following sentence starters:
  "A bubble is...."
  "A bubble is not...."
  "Here's a secret for blowing a bubble...."

Body Bubbles

What to Do

Try using your fingers to blow a bubble in the air ...

... or on the table.

Try using two hands to make bubbles.

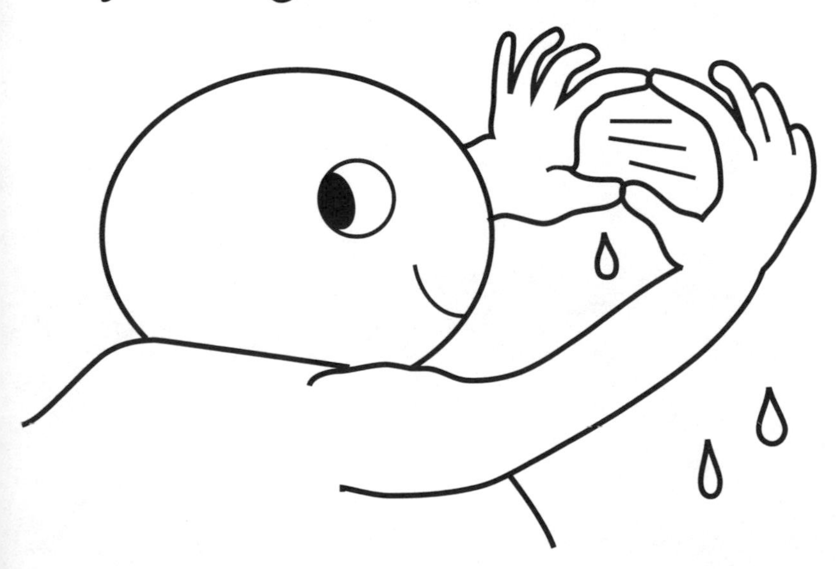

Use a straw to blow bubbles in your hand!

Body Bubbles

Questions

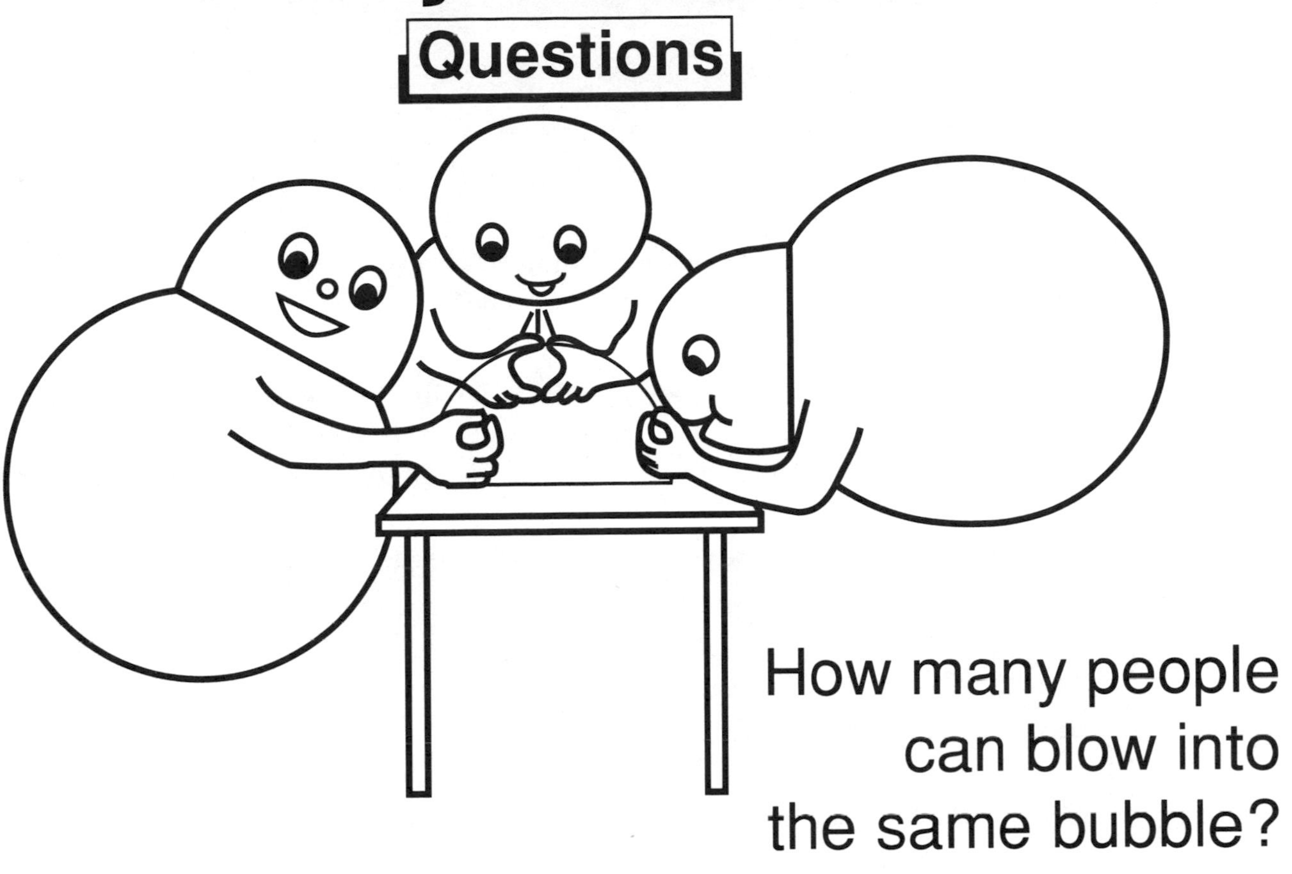

How many people can blow into the same bubble?

Can you make a line of bubbles down your arm?

Body Bubbles

Your goal is to assist the students in making their own discoveries, while keeping the activity safe, and the mess under control. Read the signs at the station so you know what the students will be investigating. If the students are non-readers, you will have to communicate the content of each sign. This is best done by giving a challenge or asking a question, rather than demonstrating how it is done. Save your demonstrations for situations when students aren't successful on their own, even with coaching.

Ask the students open-ended questions, such as: What have you discovered? Why do you think that is happening? You may also want to provide further challenges, such as: Can you think of a different way to use your hands to make a bubble? Can you use four hands to make a bubble? Six hands?

Resist the temptation to give explanations to the students.

If students get out of control, involved in creating mountains of foam or some other activity that is unrelated to the station, you might want to steer them back on task. Make sure to intervene if you see an unsafe behavior. However, keep in mind that what may appear as fooling around with bubbles can lead to some of the deepest learning experiences. Some of the greatest scientific discoveries have been made while scientists "fooled around"—the same is true of great personal discoveries.

Tips for Managing the Station

- Remind students that all surfaces touching bubbles must be wet (hands, arms, etc.).
- Use a squeegee to remove excess bubble solution from the table surface periodically.
- Throw sections of newspaper over spills on the floor.
- Refill containers of bubble solution as needed.

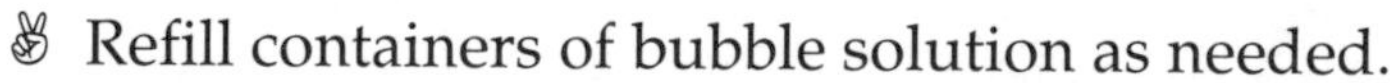

Activity 2
Bubble Shapes

Overview

By blowing clusters of bubbles on the table or in their hands, students can make bubbles that aren't round! A three-sided bubble, a four-sided bubble, a five-sided bubble, and more. By putting a wet straw into a bubble, students can blow a bubble in a bubble, or even a bubble in a bubble in a bubble. Before long, your students will be making "slinky-bubbles," "caterpillars," and many of their own clever creations.

For the youngest students, just seeing and describing the various shapes is valuable. The first step is letting them use their own words for description, such as "igloo," "crystal ball," "honeycomb," or "house." Recognizing specific geometric shapes, and referring to them by their technical names (hexagon, cube, etc.) may not occur until students are older—even if they know those shape names. It can be hard for children to see the shape of a single bubble within a cluster. To address this, some teachers laminate two- or three-dimensional shapes (triangle, square, pentagon, hexagon, cube, sphere, etc.) so younger students can search for specific bubble shapes by comparison.

While it is satisfying to hear students "discover" or name geometric shapes, keep in mind the tremendous value of their own open-ended shape explorations. Some of the most valuable learning may be occurring when we think our students are "just having fun."

For example, understanding about spheres involves more than naming their shape. Holding an undulating sphere in the hand, watching how it changes shape as it is squeezed, stretching it between two hands and seeing how the shape changes—all these activities help students understand more completely about the nature of spheres, spatial reasoning in general, and geometry. Topology, the study of how shapes and surfaces can be manipulated and changed, is a fascinating mathematical field. "Fooling around" with bubble shapes provides wonderful concrete experiences for your students, preparing them for understanding geometry and topology.

One group of students was working with animals in the science lab around the time they participated in a Bubble Festival. They were able to connect different bubble shapes they saw with structures of the animals: the compound eye of the crayfish and the fly, cells from lizard skin that they examined under the microscope. Some students tied in bubble shapes to the structure of stomates that are found on the undersides of leaves. Many of the students compared the bubble shapes they saw with the structure of geodesic domes that are common in our area.

So, enjoy your students' investigations, and remember that they may be learning something different than what you had in mind. As you circulate, think of questions to ask that make your students articulate their particular shape discoveries (What's happening to the bubble as you squeeze it?), or challenges that push your students farther (Now can you find a way to use the side of your hand to chop a bubble in two?).

One teacher had laminated paper at each station on which students drew shapes they saw with grease pencils.

What You Need for One Station

✓ 1 or 2 cottage cheese containers

Getting Ready

☞ Fill the cottage cheese containers about half-full of bubble solution.

☞ Set out the "Bubble Shapes" sign.

Special Considerations

- This activity works best when students have a large surface on which to spread out.
- Squeegee off the table periodically.

Going Further

✌ Introduce your students to the names of different polygons, polyhedrons, and/or angles that they observed in their clusters of bubbles.

✌ Have your students draw from memory the shapes that they saw.

✌ Ask your students to determine the maximum number of bubbles that touch each other at any one time.

—More **Going Further** *items are on the next page—*

Going Further

—continued—

✌ Challenge students to measure the angles between bubbles in a cluster. (If you don't want to get out the bubble solution again, take photos of bubble clusters on the day of the *Bubble Festival*. Your students can measure directly from the photo. Also, it is possible to blow bubble clusters on a sheet of acrylic plastic placed on an overhead projector so the angles between the bubbles can be seen by all.)

✌ Blow a large cluster of table bubbles, covering an entire table surface. Let your students observe the hexagon-shaped bubbles that form.

✌ Your students may discover that bubbles in large clusters most often have a hexagonal shape. As a class, investigate the many places in which the hexagon can be found—from honeycombs to the internal structure of airplane wings, skis, and molecules. Why is this a stable shape? Have your students make one or two dimensional shapes out of stir sticks and pipe cleaners (see page 103) and experiment with the relative stability of various shapes. Comparison of a square and a triangle is a good place for youngest students to focus.

✌ For older students it is possible, although difficult, to blow a cluster of bubbles in which the center bubble is shaped like a cube. It is technically hard to do this, but possible for the patient and adept bubble-blower. Ask your students to speculate on how many bubbles must be in the cluster. (A cube has six sides.)

✌ Investigate why a single bubble, when surrounded entirely by air or water, is always spherical.

✌ Ask your students to design a house made of bubbles and draw what it would look like.

✌ Can you make a bubble with four sides? Five? With straight sides?

✌ What angles do you see? What shapes do you see?

✌ Can you blow a bubble inside a bubble ... inside a bubble ... inside a ... ?

Bubble Shapes

What to Do

Blow a bunch of bubbles that touch …

… on the table …

… or in your hand.

Look at the places where bubbles meet.

Bubble Shapes

Questions For Younger Students

Can you make a bubble with straight sides?

Can you make a bubble that looks like a box?

What shapes do you see?

Can you blow a bubble
inside a bubble ...
inside a bubble ...
inside a ...

Bubble Shapes

Questions

Can you make a bubble with four sides?
Five?

Can you make a
six-sided cube
out of bubbles?

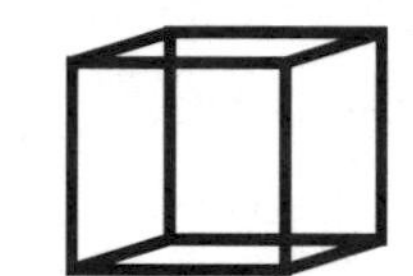

What angles
do you see?

Can you blow a bubble
inside a bubble …
inside a bubble …
inside a …

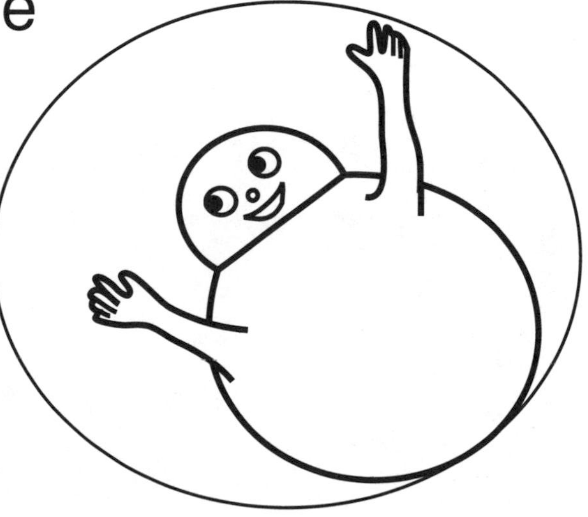

Activity Task Card for Volunteers

Bubble Shapes

Your goal is to assist the students in making their own discoveries, while keeping the activity safe, and the mess under control. Read the signs at the station so you know what the students will be investigating. If the students are non-readers, you will have to communicate the content of each sign. This is best done by giving a challenge or asking a question, rather than demonstrating how it is done. Save your demonstrations for situations when students aren't successful on their own, even with coaching.

Ask the students open-ended questions, such as: What have you discovered? Why do you think that is happening? You may also want to provide further challenges, such as: What's happening to the bubble as you squeeze it? Can you make a triangular bubble?

Resist the temptation to give explanations to the students.

If students get out of control, involved in creating mountains of foam or some other activity that is unrelated to the station, you might want to steer them back on task. Make sure to intervene if you see an unsafe behavior. However, keep in mind that what may appear as fooling around with bubbles can lead to some of the deepest learning experiences. Some of the greatest scientific discoveries have been made while scientists "fooled around"—the same is true of great personal discoveries.

Tips for Managing the Station

- Remind students that all surfaces touching bubbles must be wet (hands, straw, table), to dip the straw in solution just prior to blowing, to blow very softly, and not to inhale with their mouths on the straw.

- If you have a student who has difficulty blowing a table bubble, try blowing a bubble into his soapy hand so he can feel just how softly he needs to blow. Suggest he begin by blowing a bubble into his own palm prior to blowing a table bubble. If a student takes a quick breath in with her mouth on the straw, she is breaking the soap film at the end of the straw. Cut a diamond-shaped hole in the straw an inch or so below the blowing end to eliminate the problem.

- Keep the table surface covered with a thin coating of bubble solution. Use a squeegee to periodically remove excess bubble solution from the table.

- Throw sections of newspaper over spills on the floor.

- Refill containers of bubble solution as needed.

Activity 3
Bubble Measurement

Overview

Measure a bubble! How? It turns out that one kind of bubble is easy to measure in lots of different ways—a table bubble! Blowing gently with a soapy straw next to a soapy surface, produces a dome-shaped bubble. By inserting a wet ruler or stick in the bubble, your students can measure the height of a bubble dome or its diameter. When the bubble dome pops, a circle of soap suds is left on the table. Your students can measure its diameter, radius, or even circumference. Some students attempt to measure a bubble's volume. Students can place a pile of cubes next to a bubble of equivalent size, or they can try a more painstaking, rarely successful method of inserting soapy cubes into a bubble, one-by-one. Students have even attempted to quantify how many soapy fists can fit in a bubble dome.

Certainly the idea of blowing the biggest possible bubble is intriguing to students, and measuring is a way to verify that it is the biggest. It is less important that your students accurately measure the size of a bubble dome than that they consider the various ways they could quantify its size. As they employ different measurements, your students may discover that the height of a bubble is less than its diameter, which in turn is less than its circumference. Or they may get completely involved in seeing how many bubbles will cover the entire table surface, or in measuring the angle of a bubble intersection.

Some teachers prefer to precede Bubble Measurement with a collection of non-standard measurement experiences. Other teachers have found this to be a fine introduction to non-standard measurement.

Even if your students drift away from measuring their bubbles, they may be involved in qualitative measurements, saying such things as, "This is my biggest bubble." As you circulate between groups, you may want to guide these students into further qualitative measures by asking, "Which bubble do you estimate is bigger, your's or Yolanda's?" "How much of the desk did your bubble cover?" "How many bubbles do you think it would take to cover this desk?" "Do you think your next bubble will be bigger than this one?" "What's a quick way you could estimate the size of this bubble, so you will know if the next one is bigger?"

As with most things, it is important to remember that students will approach this activity from their own level of background and experience. While one student may focus on the concept that a bubble can be measured, another may notice a relationship between circumference and diameter. Keeping the open-ended tone to this activity will enable all students to succeed.

BUBBLE DIAMETERS
How many cubes?

30
•
•
•
7
6
5
4
3
2
1
0

Tom Lynn Sue John Anne

One teacher had her kindergarten students record their estimates of the diameter of each bubble, measure how many cubes it was, and then record their measurements. Their estimates got more and more accurate as they gained in experience estimating and then verifying the size of a bubble. Recording can be done with grease pencils on a laminated chart.

What You Need For One Station

✓ 1 or 2 cottage cheese containers

✓ A collection of standard and non-standard measuring equipment, (include enough items so that the students at your station each have at least two different tools to choose from at any time)

- Plastic-coated measuring tapes
- Metersticks, yardsticks or rulers that you don't mind getting wet
- Protractors
- About 50 unifix cubes (or other uniform, waterproof cubes —about 1 cm)
- Several pieces of string or yarn (about 1 yard or 1 meter long)
- Popsicle sticks
- Pencils
- Coffee stirrers

Optional

- Toothpicks
- Uniform-sized buttons
- Concentric rings
- Laminated circles of labeled diameters

Getting Ready

☞ Fill the 2 cottage cheese containers about half-full of bubble solution.

☞ Put the measuring equipment at the station.

☞ Set out the "Bubble Measurement" sign.

Special Considerations

- This activity works best when students can have a large surface on which to spread out.

- Squeegee off the table periodically.

- You may need to remind your students to keep their measuring tools wet so as not to pop the bubbles. Some teachers place the measuring tools in a bowl of soap solution at the station so their students have one less thing to remember.

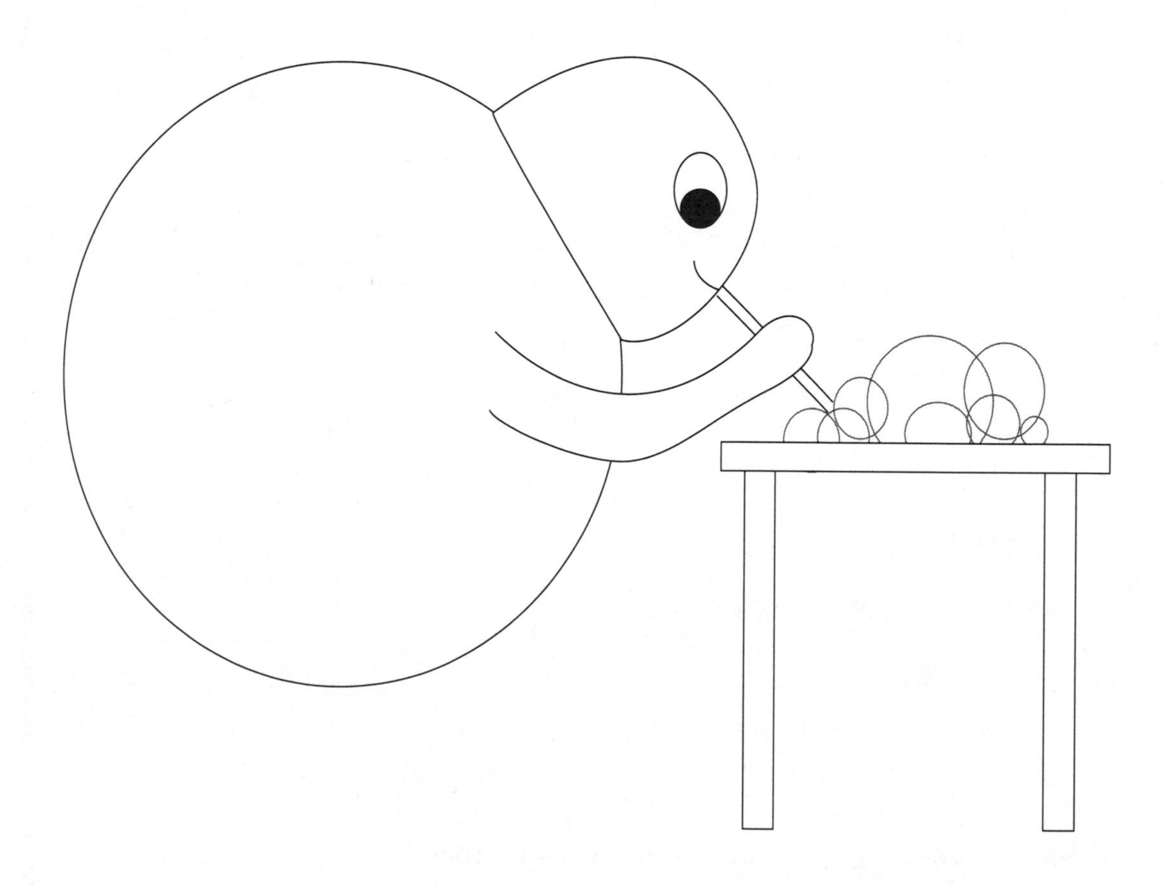

Going Further

- What other things are shaped like spheres? Draw a picture of spherical things. Write a story about a little sphere and a big sphere.

- Present more math activities which involve students with diameter, radius, and volume, shapes, polygons, and angles.

- Formalize the concepts of diameter, radius, circumference, and volume by cutting an orange in pieces, and show students what each of these words means. After students learn what diameter is, for instance, ask the students if anybody measured the diameter of her bubble, and how she did it. Make a chart of all the ways the radius of a bubble was measured, the volume of a bubble, etc.

- Discuss the advantages of using non-standard measurement techniques (such as wet cubes) and the advantages of using standard measurement techniques (such as rulers).

- Ask older students how they might determine whether a bubble dome is a perfect hemisphere. If they need help, suggest that when a bubble dome's height is equal to its radius, then that bubble dome is a perfect hemisphere. If you give your students a chance to investigate this first-hand, they will find that very large bubbles become flat on top—they are actually shorter than their radius. Can your students determine the range of bubble size at which bubble domes are perfect hemispheres?

- Calculate the volume of a bubble dome. First, assume that a bubble dome is half of a sphere. Measure the radius. One way is to put a measuring tool vertically into the bubble dome, from the table to the highest part of the dome. Another way is to measure the diameter of the ring left by a table bubble, and divide by 2. Calculate the volume of a sphere with the formula $V=\frac{4}{3}\pi$ r3 (π = 3.14) Then divide by two to determine the volume of half a sphere.

- Calculate lung volume. Have each student blow a bubble dome with three lung-fulls of air. Have them determine the volume of that bubble dome and divide by three to find the average lung volume.

- Make bubble tessellation drawings, using the sphere as a basic unit in this repeating pattern.

Bubble Measurement

What to Do

Blow a table bubble. When it pops, it will leave a circle on the table—a bubble print!

The distance across the widest part of the bubble print is called the diameter.

Try measuring the diameter of your bubble with some of the measuring tools at the station, or even with your hand.

Bubble Measurement

What to Do For Younger Students

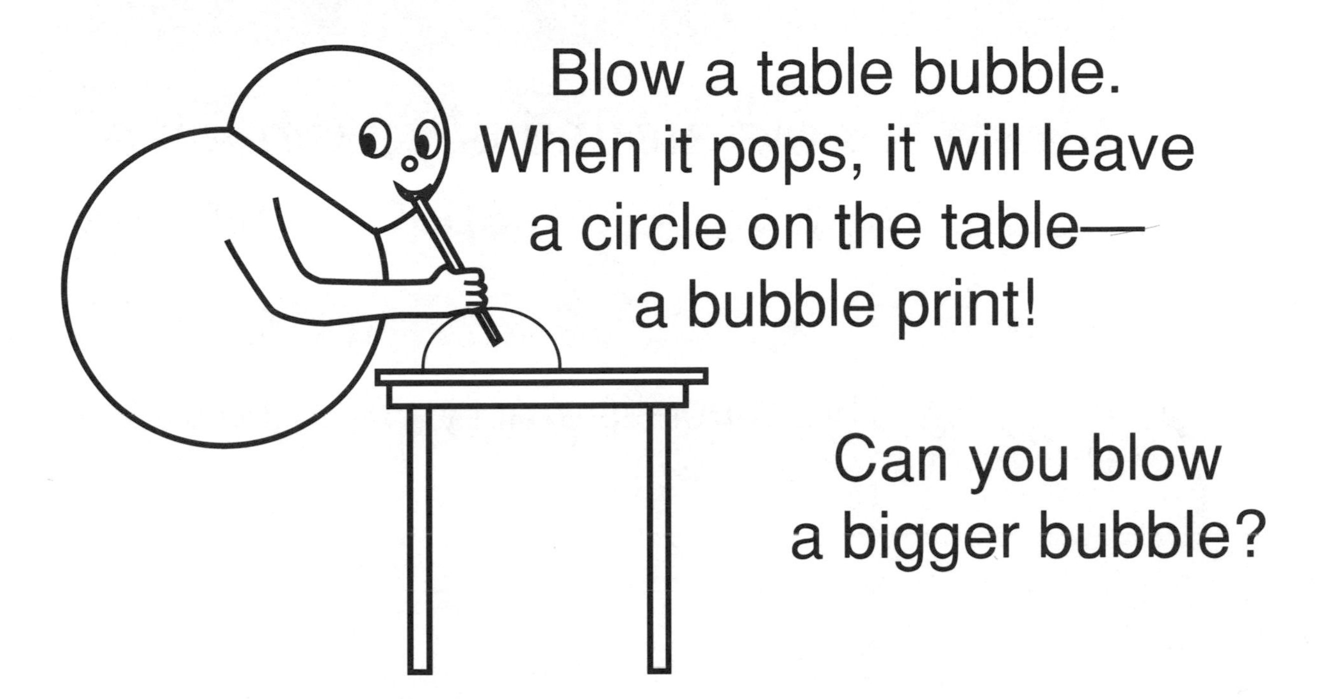

Blow a table bubble.
When it pops, it will leave
a circle on the table—
a bubble print!

Can you blow
a bigger bubble?

How many bubbles will it take
to cover the table?

Bubble Measurement

Questions

How tall or big around is your bubble?
How big is the space inside it?

Try three different ways
of measuring your bubble.

Activity Task Card for Volunteers

Bubble Measurement

Your goal is to assist the students in making their own discoveries, while keeping the activity safe, and the mess under control. Read the signs at the station so you know what the students will be investigating. If the students are non-readers, you will have to communicate the content of each sign. This is best done by giving a challenge or asking a question, rather than demonstrating how it is done. Save your demonstrations for situations when students aren't successful on their own, even with coaching.

Ask the students open-ended questions, such as: What have you discovered? Why do you think that is happening? You may also want to provide further challenges, such as: Can you think of another way to measure a bubble? What's the biggest bubble you can make? The smallest?

Resist the temptation to give explanations to the students.

If students get out of control, involved in creating mountains of foam or some other activity that is unrelated to the station, you might want to steer them back on task. Make sure to intervene if you see an unsafe behavior. However, keep in mind that what may appear as fooling around with bubbles can lead to some of the deepest learning experiences. Some of the greatest scientific discoveries have been made while scientists "fooled around"—the same is true of great personal discoveries.

Tips for Managing the Station

- Remind students that all surfaces touching bubbles must be wet (hands, straw, table, measuring implements), to dip the straw in solution just prior to blowing, to blow very softly, and not to inhale with their mouths on the straw.

- If you have a student who has great difficulty blowing a table bubble, try blowing a bubble into his soapy hand so he can feel just how softly he needs to blow. Suggest that he begin by blowing a bubble into his own palm prior to blowing a table bubble. If a student takes a quick breath in with her mouth on the straw, she is breaking the soap film at the end of the straw. Cut a diamond-shaped hole in the straw an inch or so below the blowing end to eliminate the problem.

- Keep the table surface covered with a thin coating of bubble solution. Use a squeegee to periodically remove excess bubble solution from the table.

- Throw sections of newspaper over spills on the floor. Refill containers of bubble solution as needed.

Activity 4
Bubble Technology

Overview

Technology involves the use of science to create something practical. In this activity, your students experiment to discover what objects can be used to blow bubbles, which make little bubbles, and which make big bubbles. By gathering this information, they are doing a kind of technical research. You may want to give them a chance after the *Bubble Festival* to use their research to design and draw specialized bubble blowers.

What You Need for One Station

✓ 2 dish pans

✓ At least 15 pieces of "junk" to use for bubble-makers, such as: strainers, small tin cans, protractors, paper, mason jar lids, string, drinking straws, tea ball, rubber stoppers with holes, flower pots, funnels, strawberry baskets, plastic rings from a six-pack, medicine droppers, a length of rope, paper cups, styrofoam cups, various mesh sizes of screen, different size washers, rubber bands, toilet paper, paper towel rolls, aluminum foil, wire of different gauges, and springs.

Note: You may want to avoid items such as tubing (students tend to put their mouths on it) and turkey basters (squirt fights anyone?). Unfortunately, these items are really neat bubble-makers. Be advised that some metal items will rust. If you choose to use metal items that are not enameled or made of stainless steel, make sure they are rinsed and dried after use.

✓ Although most of the materials you choose should be potentially successful bubble-makers, it is important to include some that won't work (e.g., cups or spoons). Part of the activity involves categorizing objects according to whether or not they make bubbles.

Getting Ready

☞ Fill the dish pans about half-full with bubble solution, enough so objects can be completely dipped.

☞ Put the "junk" out on the station.

Special Considerations

- Foamy soap solution makes it harder to make bubbles. You may want to ask your students not to jiggle the objects in the bubble solution—too much jiggling makes foam. Skim the foam off the surface of the bubble solution periodically.

- Remind your students to leave the items on the table, not in the solution. Some teachers put the items on a tray, and ask the students to put them back on the tray when they are done.

- There are two methods for making a bubble with the various objects: blowing through the object or waving it in the air. Waving is the messier of the two methods as bubble solution flies off the item when it is waved. Teachers who want to limit the mess as much as possible could ask their students not to wave, or have a special waving area on a drop cloth.

Going Further

- Ask your students to choose one object in the classroom, or bring an object from home, which will work to make bubbles. Ask them to predict whether it will make big bubbles or small bubbles. Allow each student to test his object, one at a time, in front of the class. Have the rest of the students predict whether they think it will work, and if it didn't, analyze why not.

- Have students create drawings of bubble-makers for specialized uses, such as a bubble-maker that makes foam, one that doesn't need to be dipped into soap solution, one that makes large, detachable bubbles, or one that makes five bubbles at a time.

- Have your students combine materials to make more complicated bubble-blowing contraptions and machines.

- Ask your students to imagine that they are stranded on a strange planet and they must create a machine to make a bubble that is big enough to float away on. Have them draw and write about their machines.

- Use this station as a starting point for a unit on inventions.

- Provide your students with other engineering challenges. Have them create an egg-cracking machine, an automatic ice cream scoop, a jelly bean factory, or a machine of their choice. Explain that technology involves the use of science to create something practical. Ask students to identify experiments that would give them information important to the design of their machine. (For example, how hard you need to hit an eggshell to crack it, or how long jelly beans take to dry.)

Bubble Technology

What to Do

Choose a tool.
Try to make a bubble with it.

Decide whether it makes a big bubble,
a small bubble,
or no bubble at all.

Bubble Technology

Questions

What's
the same about
all the tools that
make bubbles?

Can you turn a non-bubble blower
into a bubble blower?

Activity Task Card for Volunteers

Bubble Technology

Your goal is to assist the students in making their own discoveries, while keeping the activity safe, and the mess under control. Read the signs at the station so you know what the students will be investigating. If the students are non-readers, you will have to communicate the content of each sign. This is best done by giving a challenge or asking a question, rather than demonstrating how it is done. Save your demonstrations for situations when students aren't successful on their own, even with coaching.

Ask the students open-ended questions, such as: What have you discovered? Why do you think that is happening? You may also want to provide further challenges, such as: Can you use that object in a different way to make bubbles?

Resist the temptation to give explanations to the students.

If students get out of control, involved in creating mountains of foam or some other activity that is unrelated to the station, you might want to steer them back on task. Make sure to intervene if you see an unsafe behavior. However, keep in mind that what may appear as fooling around with bubbles can lead to some of the deepest learning experiences. Some of the greatest scientific discoveries have been made while scientists "fooled around"—the same is true of great personal discoveries.

Tips for Managing the Station

- Discourage students from putting their mouths directly on the objects. Blowing through the object from a distance of several inches or waving the object through the air works more effectively.

- Foamy soap solution interferes with this activity. Skim the foam off the surface of the bubble solution periodically. You may want to ask your students not to jiggle their objects in the bubble solutions as too much of this is what makes the foam.

- Refill tubs of bubble solution as needed. There should be enough bubble solution in the tub to completely cover the bubble-making objects.

- Keep the objects on the table, not in the solution. This will make them easier to find, and the tubs clearer for complete dipping.

- Use a squeegee to periodically remove excess bubble solution from the table. Throw sections of newspaper over spills on the floor.

Activity 5

Bubble Colors

Overview

We've all enjoyed the beautiful colors swirling around on bubbles. In this activity, students get a chance to carefully observe those colors and patterns. They can use their observations to tell how old a bubble is and even to predict when it will pop!

Students blow table bubbles and put bubble "homes" around them. The homes protect bubbles from air currents and reflect more white light onto the bubble, making it easier to see the colors. The dark background on the station surface provides contrast. The result is very striking!

When a bubble is first blown, students will observe wildly swirling colors on the surface of the bubble. As the bubble settles down, students will notice rings or bands of color on the bubble. (Blowing lightly on the surface of the bubble will produce wild swirling again.) As the bubble gets thinner, they will observe bands moving outward in a certain repeating sequence of colors. Just before the bubble pops, there will be a characteristic change in color and pattern that even the youngest students can notice.

This activity requires a certain level of focus. It works best if just one bubble is blown at a station, and then observed by all. Some groups become completely absorbed by the activity. Others get somewhat "antsy" after they have watched several bubbles. Depending on the attention span and excitement level of your students, you may want to be prepared to send these students to a dry area with crayons and paper to draw what they saw.

What You Need for One Station

✓ 1 black or brown plastic trash bag (about 2' by 3')

✓ 8 pieces of tagboard (approximately 8.5" x 11")

✓ 1 roll masking tape

✓ 1 or 2 cottage cheese containers

✓ 1 pair scissors

Getting Ready

☞ You need a dark surface for this station. If your table top or counter is dark-colored, you may use it as is. If it is light-colored, cut the trash bag along one side and the bottom crease so that it opens out in one flat piece. Use it to cover the station surface as follows: Sprinkle a few drops of water on the table or counter surface first. Then lay the trash bag down, black side up, and smooth it down. The slight dampness will make the plastic adhere to the surface.

☞ Make two bubble "homes." Staple the tagboard together, end-to-end, until you have made a circular enclosure. Set the two bubble homes at the station.

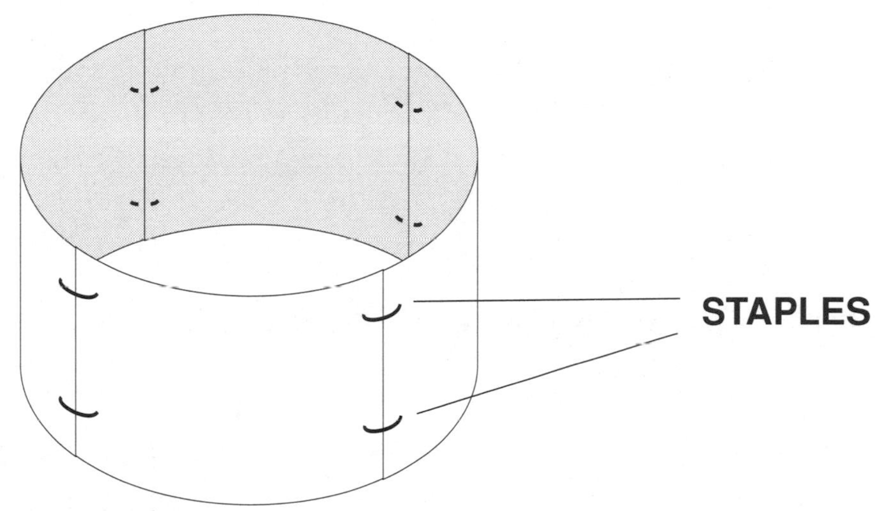

☞ Fill each cottage cheese container about half-full with bubble solution.

☞ Set out the "Bubble Colors" sign.

Special Considerations

- This activity works best when small groups blow and observe one bubble at a time until it pops. It is hard for students to refrain from blowing lots of bubbles, especially if this is an early encounter with bubbles. Schedule this activity later in your sequence of sessions for greatest success.

Going Further

- Have students use crayons to draw a picture of a bubble and the colors they remember seeing.

- Have your students draw pictures of a bubble at various life stages. Or as a class project, take photographs of a bubble at these different stages.

- Have young students dictate stories about what they saw. Have older students write a letter instructing someone on how they can use color to learn about a bubble.

- Tell the students that the colors we see on bubbles come from white light. Explain that white light is made of light of many colors. With older students, you might want to explain that as the thickness of the bubble wall changes, so do the colors we see on that part of the bubble wall. See page 132 for more on "Light and Color."

- Use a prism or diffraction grating to show how white light is made up of light of different colors.

- Have students hypothesize about the difference between the colors you can see on a bubble in a bubble home versus those that can be seen on a bubble not in a bubble home. Have them hypothesize why.

- Conduct an experiment to see if the color of the bubble homes makes a difference to the colors you see on a bubble. Set up a line of bubbles, each blown in a bubble home of a different-colored tagboard, and observe the differences.

Bubble Colors

What to Do

Blow just one table bubble at this station.

Put a bubble "home" around it.

Watch the colors and patterns on the bubble until it pops.

Bubble Colors

Questions

What colors and patterns do you see?

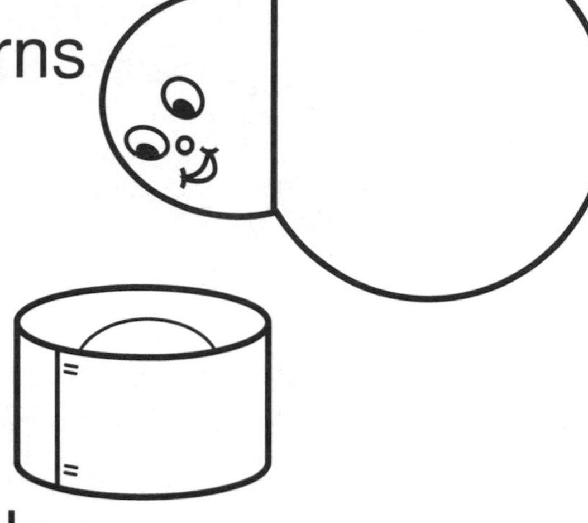

What happens if you blow lightly on a bubble?

Can you tell how old a bubble is by its colors?

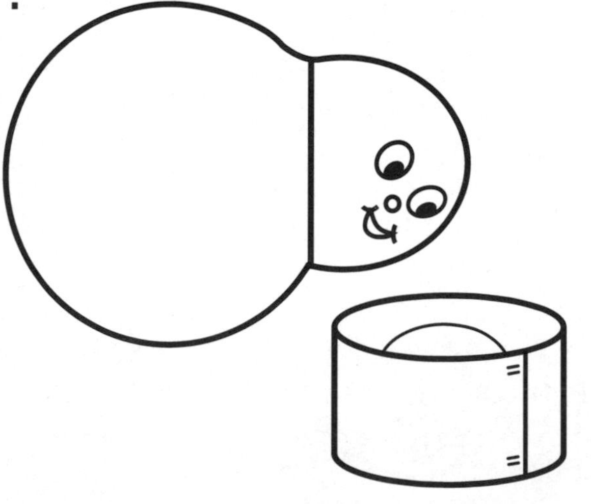

Can you tell exactly when a bubble will pop?

Five, four, three, two, one ...

Activity Task Card for Volunteers

Bubble Colors

Your goal is to assist the students in making their own discoveries, while keeping the activity safe, and the mess under control. Read the signs at the station so you know what the students will be investigating. If the students are non-readers, you will have to communicate the content of each sign. This is best done by giving a challenge or asking a question, rather than demonstrating how it is done. Save your demonstrations for situations when students aren't successful on their own, even with coaching.

Ask the students open-ended questions, such as: What have you discovered? Why do you think that is happening? You may want to provide further challenges, such as: Is there an order to the colored rings? Can you keep a bubble from popping by blowing on it?

Resist the temptation to give explanations to the students.

If students get out of control, you might want to steer them back on task. Make sure to intervene if you see an unsafe behavior. However, keep in mind that what may appear as fooling around with bubbles can lead to some of the deepest learning experiences. Some of the greatest scientific discoveries have been made while scientists "fooled around."

Tips for Managing the Station

- Remind students that all surfaces touching bubbles must be wet (hands, straw, table), to dip the straw in solution just prior to blowing, to blow very softly, and not to inhale with their mouths on the straw.

- If you have a student who has great difficulty blowing a table bubble, try blowing a bubble into his soapy hand so he can feel just how softly he needs to blow. Suggest that he begin by blowing a bubble into his own palm prior to blowing a table bubble. If a student takes a quick breath in with her mouth on the straw, she is breaking the soap film at the end of the straw. Cut a diamond-shaped hole in the straw an inch or so below the blowing end to eliminate the problem.

- This activity works best when small groups blow and observe one bubble at a time until it pops. You may want to organize the students at the station so they take turns being the one to blow the bubble, while the others call out the colors they see. Remind students to use the white bubble "homes." These will reflect more light, allow students to see more colors, and protect the bubble.

- Keep the surface of the black plastic covered with a thin coating of bubble solution. Use a squeegee to periodically remove excess bubble solution from the table. Throw sections of newspaper over spills on the floor, and refill containers of bubble solution as needed.

Activity 6
Bubble Windows

Overview

At this station, students make sheets of soap film by slowly raising straws and a loop of string out of a tub of bubble solution. Surface tension within the bubble film causes the sides of the apparatus to pull together. Students experiment with the flexible bubble film by blowing on it, pulling it through the air, waving it gently, or even twisting it. Students discover how to make one window intersect another, poke things through the windows without popping them, and how the world looks through a bubble window.

As an optional challenge, you can have students use straws to blow through the soap film of a bubble window. Done in a certain way, students can cause small bubbles to form from the soap film and float away. Some students have even used their bubble windows as trampolines on which these smaller bubbles bounce!!

What You Need for One Station

✓ 2-3 dish pans

✓ At least 6 yards (or meters) of absorbent cotton string or twine. Be sure the string is not too thin or tightly twisted. The more bubble solution it can absorb, the better.

✓ 12 drinking straws

Getting Ready

☞ Make six bubble windows as follows: Cut the cotton string into six pieces, each approximately one yard (1 meter) in length.
(Note: Longer string will make bigger windows. A few bigger windows can be fun, especially for older students. Keep in mind that the bigger the windows get, the harder they are to manage.)
Thread each piece of string through two straws. Tie the ends of the string together. Pull the straws around to opposite sides of the loop. Shift the knot into one of the straws so that it is out of the way. You now have a loop with two straws on it for handles. Repeat these directions to make the other five bubble windows. *(See illustration on next page.)*

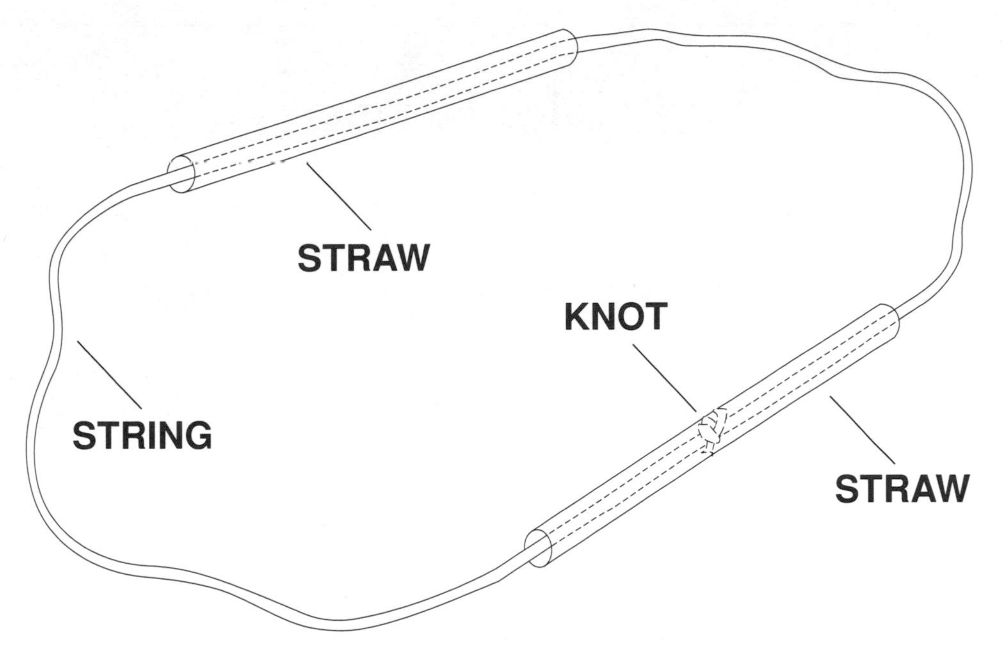

☞ Fill the dish pans about a third full of bubble solution. Put two or three string loops in each dish pan.

☞ Set out the "Bubble Windows" sign.

Special Considerations

- This activity wins the prize for being the messiest! Nevertheless, nearly every teacher who has presented it to her students ranked it as one of the best, even after clean-up. If possible, set it up on a drop cloth. Teachers who want to limit the mess as much as possible ask their students not to wave, but this is very difficult to enforce, as waving produces the most fantastic bubbles. Because of the waving, this activity needs more space than others.

- Suggest to your students that they do everything in slow motion at this station. This will help them be more successful, and nicely calms the excitement level.

- Foamy soap solution interferes with this activity. You may want to ask your students not to jiggle their windows in the bubble solution too much as jiggling makes the foam. Skim the foam off the surface of the bubble solution periodically.

- Students who hold the windows over their heads and then look up through them will have bubble solution drip into their eyes. If you see this happening, you may want to caution students about it.

Going Further

- The sides of the bubble windows pull towards the center because of the surface tension of the bubble solution. Give your students some first-hand experience with surface tension with the following experiment.

  Give each student an eye dropper and a penny. Ask them to predict how many drops of water will fit on the penny without spilling. Distribute dishes of water and have them find out. Ask the students if they can see how the water behaves as if it were covered with a skin.

  Explain that this effect is called surface tension. Water molecules at the surface of water are more attracted to each other than to the air; it is as if they stick together. This attraction of water to itself is what causes the sides of the bubble windows to pull towards the center.

  Have your students repeat this after wiping some dish soap on the penny. How does the soap affect the surface tension of water?

- Using a bright light source, project the bubble windows on to a screen and watch as the solution runs down.

- Have your students write about color and movement.

- Measure the windows in your classroom.

- Have your students do some fantasy writing: "I looked through my bubbly, hazy, window, and I saw ... "

- Determine whether blowing on or waving a window makes a bigger bubble.

Bubble Windows

What to Do

Hold on to the straws, and dip the whole loop of string into the solution.

Slowly lift it out!

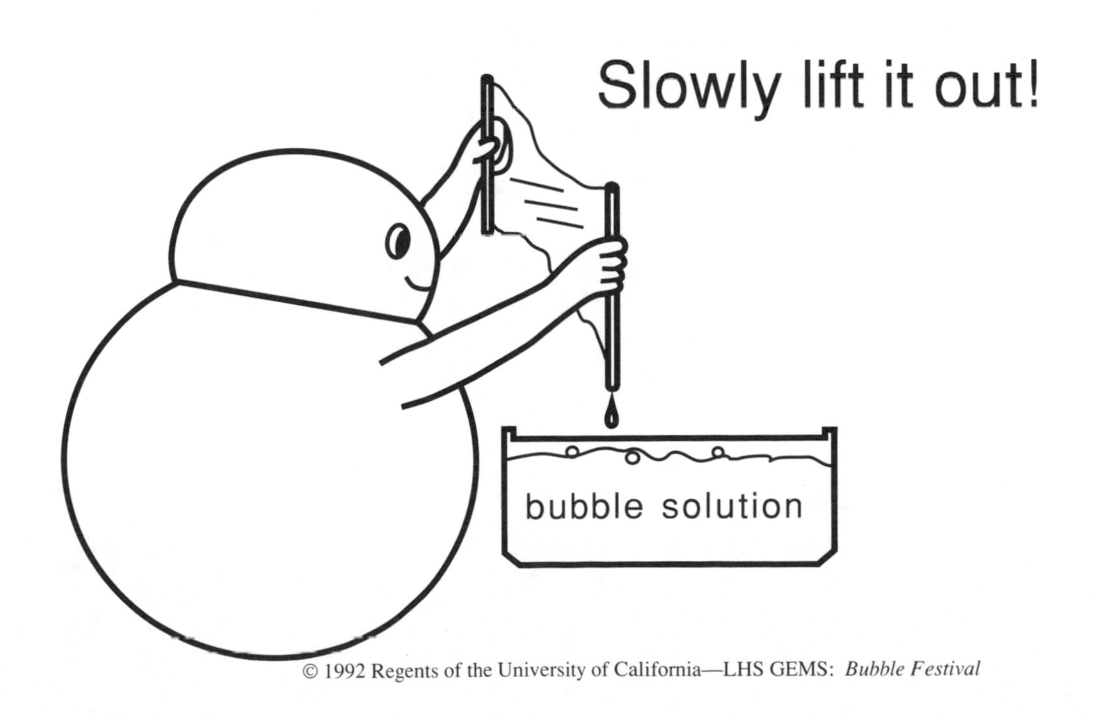

Bubble Windows

Questions

What happens when you pull your window through the air?

Can you poke things through without popping it?

Can a friend put another window through yours?

How many ways can you make a butterfly?

Activity Task Card for Volunteers

Bubble Windows

Your goal is to assist the students in making their own discoveries, while keeping the activity safe, and the mess under control. Read the signs at the station so you know what the students will be investigating. If the students are non-readers, you will have to communicate the content of each sign. This is best done by giving a challenge or asking a question, rather than demonstrating how it is done. Save your demonstrations for situations when students aren't successful on their own, even with coaching.

Ask the students open-ended questions, such as: What have you discovered? Why do you think that is happening? You may also want to provide further challenges, such as: Hold a bubble window vertically for a minute or two. Can you notice a pattern to the colors you see?

Resist the temptation to give explanations to the students.

If students get out of control, involved in creating mountains of foam or some other activity that is unrelated to the station, you might want to steer them back on task. Make sure to intervene if you see an unsafe behavior. However, keep in mind that what may appear as fooling around with bubbles can lead to some of the deepest learning experiences. Some of the greatest scientific discoveries have been made while scientists "fooled around."

Tips for Managing the Station

- If students are having difficulty making bubble windows, remind them that all surfaces touching bubbles must be wet (hands, strings, and straws).

- Suggest that students do everything in slow motion at this station. This will help them be more successful, and it nicely moderates the excitement level.

- Lifting the string and straw out of the bubble solution at an angle, rather than straight up, is more effective.

- Discourage students from walking blindly backwards to make bubbles. Encourage them to wave their windows through the air using an up-down motion while standing still instead.

- Foamy soap solution interferes with this activity. Skim the foam off the surface of the bubble solution periodically. You may want to ask your students not to jiggle their windows in the bubble solution as too much of this makes foam.

- Refill tubs of bubble solution as needed. There should be enough bubble solution in the tub to completely cover the string-and-straws frame.

- Use a squeegee to periodically remove excess bubble solution from the table. Throw sections of newspaper over spills on the floor.

Activity 7
Bubble Walls

Overview

Students make giant sheets of soap film by slowly raising a dowel out of a trough of bubble solution. By blowing and poking at these bubble walls, students can investigate the properties of soap films, seeing how far the films stretch and twist. Surface tension within the film causes the strings on the dowel to pull together.

If you set up two troughs with their long sides only a few inches apart, you can add a challenge that isn't listed on the station sign. By blowing, can the students make a soap film "tunnel" connecting two bubble walls?

As an optional challenge, you can have students use straws to blow through the soap film of a bubble wall. Done in a certain way, students can cause small bubbles to form from the soap film and float away.

What You Need for One Station

✓ 1 roll of duct tape

✓ 2 wallpaper troughs (About 30" x 6" x 6"— these inexpensive, plastic troughs are available at paint or wallpaper supply stores.)

✓ 2 wooden dowels, about ½ inch (1.5 cm) in diameter, about 29" (73 cm) long (or whatever length fits into your troughs)

✓ About 4 yards (360 cm) of absorbent cotton twine

✓ 4 large, heavy stainless steel washers or weights about 1½ inches (4 cm) in diameter

Getting Ready

☞ On the day before the *Bubble Festival*, cut two dowels so that they fit into the troughs, reaching almost to the ends. Tie a one-yard piece of string securely to the ends of the dowels. To the other end of each piece of string, tie a washer or weight.

☞ On the day of the activity, place the two troughs at the station with their long sides touching.

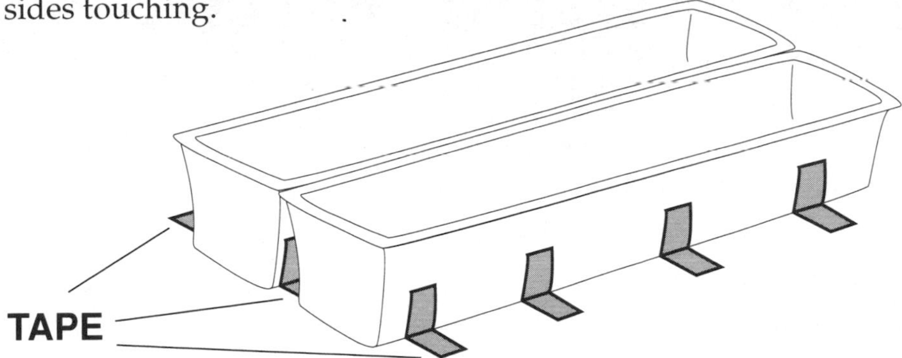

(If the station is a counter top or a very wide table, students may not be able to use both troughs in this position. In that case, place the troughs separately wherever convenient.) Tape the two wallpaper troughs to the station surface with enough duct tape to keep them steady. Fill the troughs about one-third full of bubble solution, and put a dowel-string apparatus in each trough.

☞ Set out the "Bubble Walls" sign.

Special Considerations

- This activity can get very wet. If possible, lay a drop cloth in front of the troughs of solution. Provide enough space between this and other activities so kids have room to move easily.

- Mention to your students that the string has to be inside the trough. A wall will not form unless the soap solution has a continuous area on which to form.

- Foamy soap solution interferes with this activity. You may want to ask your students not to jiggle the walls in the bubble solution as jiggling makes foam. Periodically skim the foam off the surface of the bubble solution.

Going Further

✌ Ask your students to fantasize about what's beyond the bubble wall. Have them draw or write their fantasy.

✌ Attach a string in the shape of an upside-down lollipop to the dowel, being careful not to allow it to tangle with the side strings. Make a bubble wall with this string incorporated as part of the wall. Using a dry finger, pop the soap film in the small circle. See what happens!

Bubble Walls

What to Do

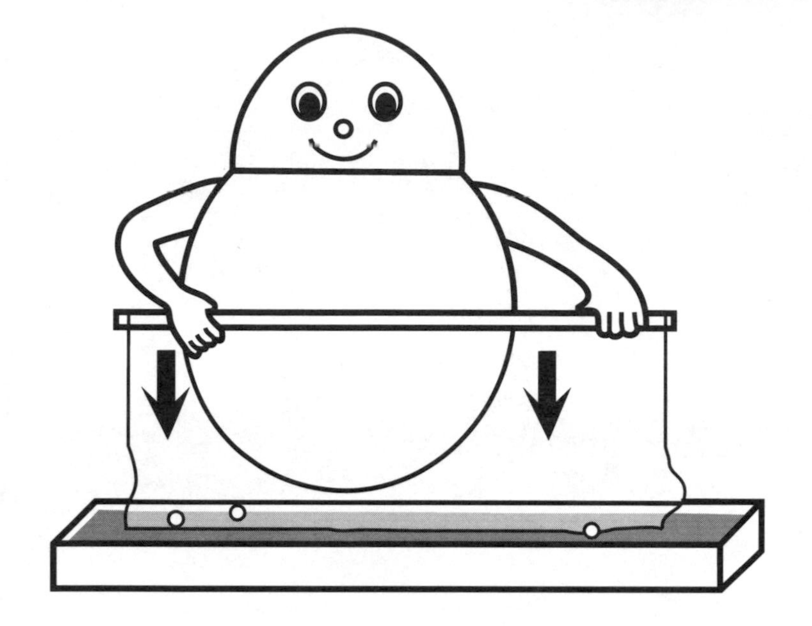

Dip the stick and string into the solution …

… and slowly lift the stick out!

Bubble Walls

Questions

What do things look like through the bubble wall?

Can you make the wall sway?

Can you shake hands through the window?

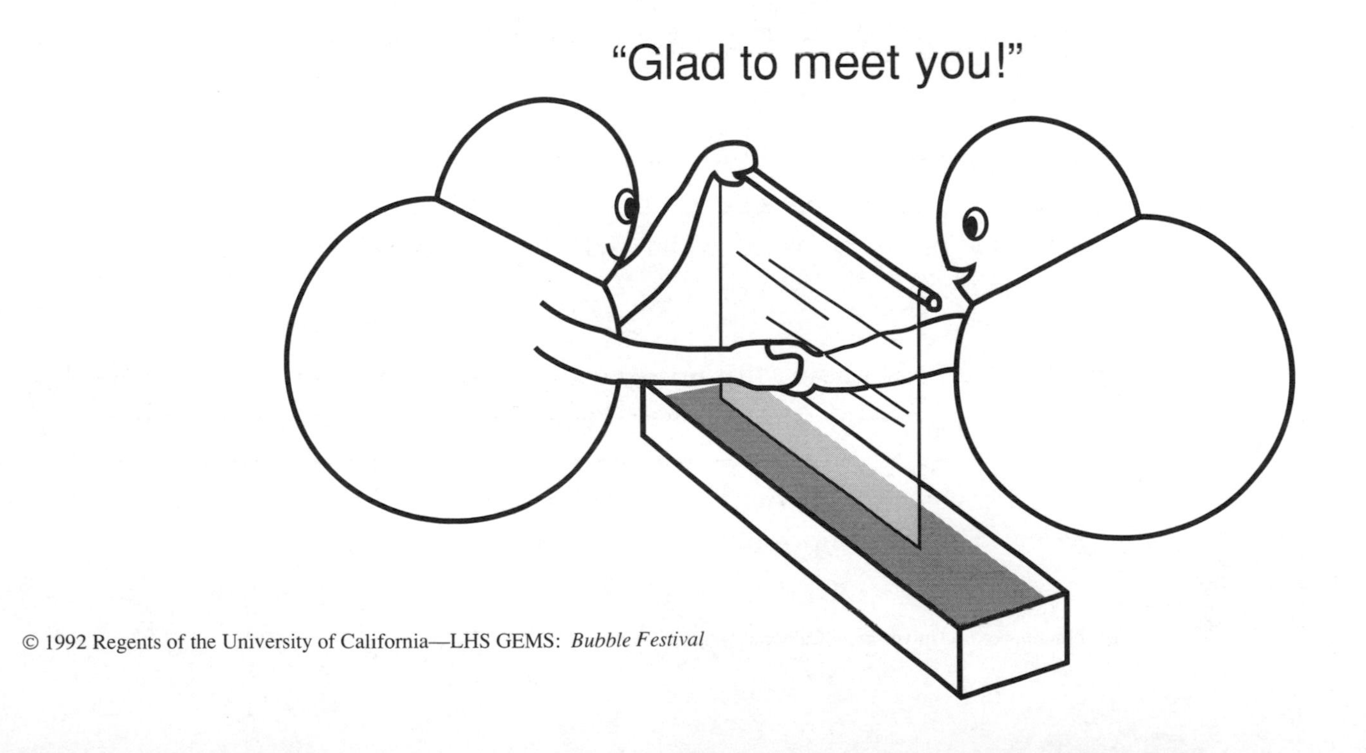

Activity Task Card for Volunteers

Bubble Walls

Your goal is to assist the students in making their own discoveries, while keeping the activity safe, and the mess under control. Read the signs at the station so you know what the students will be investigating. If the students are non-readers, you will have to communicate the content of each sign. This is best done by giving a challenge or asking a question, rather than demonstrating how it is done. Save your demonstrations for situations when students aren't successful on their own, even with coaching.

Ask the students open-ended questions, such as: What have you discovered? Why do you think that is happening? You may also want to provide further challenges, such as: Can you blow a small bubble from a bubble wall? Can you make a bubble wall wiggle?

Resist the temptation to give explanations to the students.

If students get out of control, involved in creating mountains of foam or some other activity that is unrelated to the station, you might want to steer them back on task. Make sure to intervene if you see an unsafe behavior. However, keep in mind that what may appear as fooling around with bubbles can lead to some of the deepest learning experiences. Some of the greatest scientific discoveries have been made while scientists "fooled around."

Tips for Managing the Station

- If students are having difficulty making bubble walls, mention that the string has to be inside the trough. Also remind them that all surfaces touching bubbles must be wet (hands, strings, and dowels).

- Suggest that students do everything in slow motion at this station. This will help them be more successful, and it nicely moderates the excitement level.

- Foamy soap solution interferes with this activity. Skim the foam off the surface of the bubble solution periodically. You may want to ask your students not to jiggle the walls in the bubble solutions as too much of this makes foam.

- Refill tubs of bubble solution as needed. There should be enough bubble solution in the tub to completely cover the dowel-and-string frame.

- Use a squeegee to periodically remove excess bubble solution from the table. Throw sections of newspaper over spills on the floor.

Activity 8
Bubble Foam

Overview
Younger students enjoy stirring up bubble solution and experimenting with the resulting foam. They discover that the suds are made of tiny bubbles. Can they count or estimate how many bubbles in a handful of foam? What does foam look like through a magnifying lens? Which makes better foam, a whisk or a beater? Can bubble foam be compressed? Can they fill a bucket with foam?

Both the management challenges of this station and its relatively simple content make this activity unlikely to be appropriate for older students.

What You Need for One Station
✓ 2 dish pans

✓ 2 whisks

✓ 2 egg beaters

✓ 2 empty plastic cups

✓ 2 empty buckets

✓ 4 small plastic magnifying lenses

Getting Ready
☞ Fill each dish pan with about two inches of soap solution. Put a whisk and an egg beater in each dish pan.

☞ Put hand lenses at the station and cups on station table. Place empty buckets next to the table.

☞ Set out the "Bubble Foam" signs.

Special Considerations
- Don't allow kids to put foam on their faces.
- Keep your eye on the transfer of foam to the bucket.

Going Further

- Allow pairs of students to investigate dabs of shaving cream with hand lenses. What do the tiny bubbles look like?

- Have the students take turns whipping cream or egg whites. Talk about what's inside of bubbles. Add flavoring and sweetener to the whipped cream, or cook the sweetened, whipped egg whites to make meringues to eat.

- In a ziplock bag, add an envelope of dry baker's yeast, a teaspoon of sugar, and warm water (approximately 50° F). Seal the bag, removing as much air as you can. Set the sealed ziplock bag in a styrofoam cup about half-full of warm water. After about 10 minutes, the yeast will have produced a foamy mass. These bubbles of carbon dioxide gas are what make bread rise.

- Bring in slices of bread and have students observe the tiny bubbles with a hand lens.

- Ask your students to bring in things from home that have bubbles in them. They may bring carbonated drinks, foam rubber, baked goods, cheese puffs, packing material, or other items. Talk about what's inside the bubbles. *(In most cases, it will be air or carbon dioxide.)*

- Have your students write a foam poem.

- Bubble Artwork. Mix tempera paint in a wide-mouthed container with bubble solution—mostly bubble solution with a squirt of paint. Have students use a straw to blow into the solution until foamy bubbles rises over the top of the container without spilling. With white construction paper, or other absorbent paper, have the students lay the paper on top of the foams. As the bubbles pop, they will leave a print on the paper. Have them carefully remove the paper and set it flat to dry. Add more paint if the prints are too light. Add more bubble solution if it's difficult to make foam.

- Collect some foam in berry baskets, or in a bucket. See how long it lasts.

- Use bubble foam to insulate an ice cube. Which melts faster in a bucket of foam: an ice cube in a cup or an ice cube in a cup? Discuss how our homes are insulated with materials that trap air (fiberglass) or how styrofoam is a good insulating material (it's filled with bubbles of air).

Bubble Foam

What to Do

Make lots of foam!

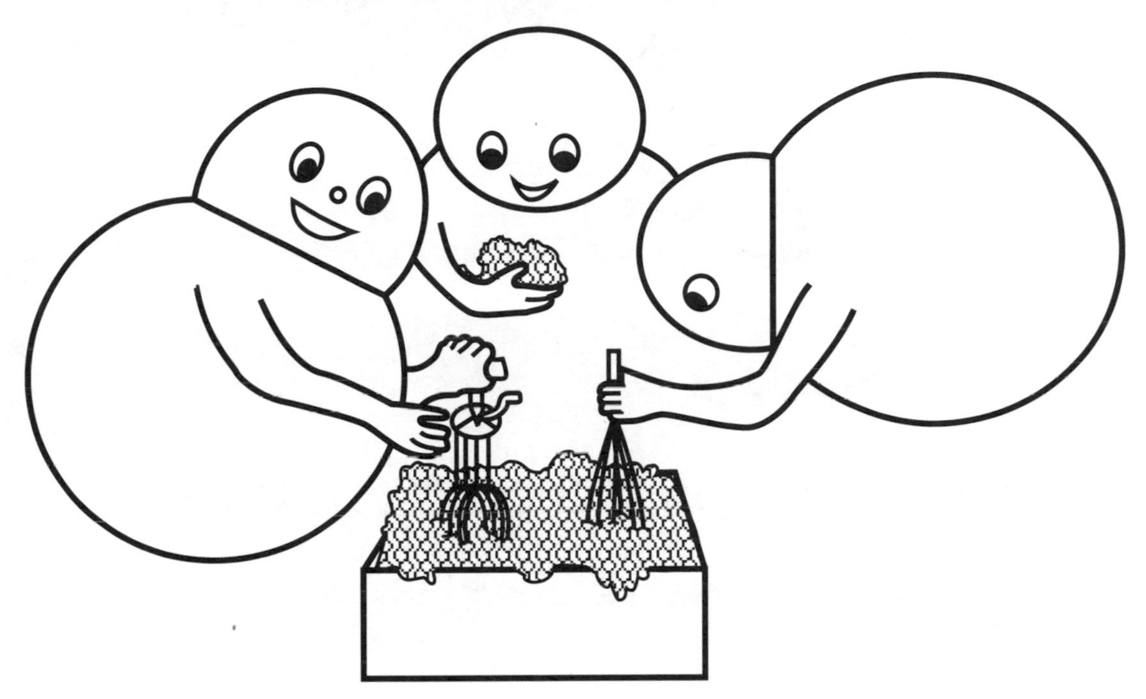

Can you fill a bucket with foam?

Bubble Foam

Questions

What is foam?

Can you count the tiny bubbles?

Put your straw in
and blow.

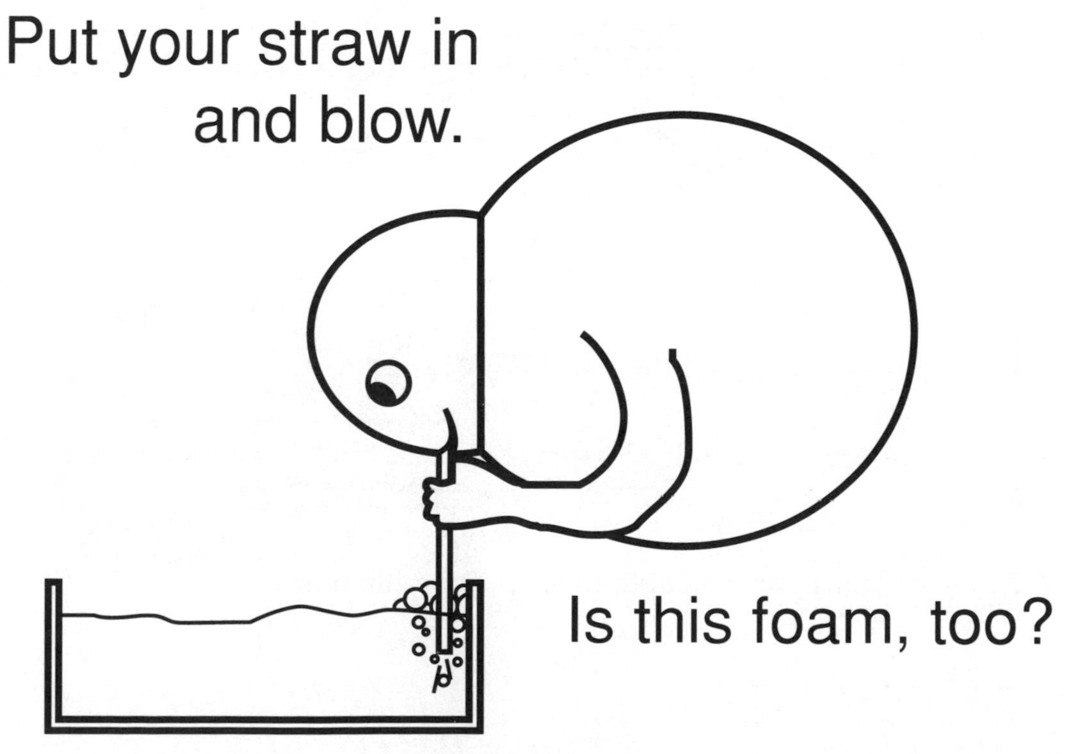

Is this foam, too?

Bubble Foam

Your goal is to assist the students in making their own discoveries, while keeping the activity safe, and the mess under control. Read the signs at the station so you know what the students will be investigating. If the students are non-readers, you will have to communicate the content of each sign. This is best done by giving a challenge or asking a question, rather than demonstrating how it is done. Save your demonstrations for situations when students aren't successful on their own, even with coaching.

Ask the students open-ended questions, such as: What have you discovered? Why do you think that is happening? You may also want to provide further challenges, such as: Can you make foam with smaller bubbles? With bigger bubbles? How does the foam change if you beat faster? Can you count the bubbles?

Resist the temptation to give explanations to the students.

If students get out of control, involved in using egg beaters to make music or some other activity that is unrelated to the station, you might want to steer them back on task. Make sure to intervene if you see an unsafe behavior. However, keep in mind that what may appear as "fooling around" with bubbles can lead to some of the deepest learning experiences. Some of the greatest scientific discoveries have been made while scientists "fooled around" — the same is true of great personal discoveries.

Tips for Managing the Station

✌ Keep your eye on the transfer of foam to bucket. If students are not doing this themselves, then you will want to periodically remove foam from the solution by scooping into the bucket. If the bucket gets too full, dump foam in a nearby, plastic-lined garbage can. After several hours, the foam will "disappear."

✌ Don't allow the students to put foam on their faces.

✌ Refill tubs of bubble solution as needed.

✌ Use a squeegee to periodically remove excess bubble solution from the table.

✌ Throw sections of newspaper over spills on the floor.

Activity 9
Bubble Skeletons

Overview

Students dip three-dimensional "skeleton" shapes into bubble solution. Soap film clings to the skeletons, creating beautiful and interesting geometrical shapes. Students watch what happens to the soap film when they pop one section at a time. An extra challenge is to find a way to pop and re-dip the cube shape to create a *little* bubble-cube in the center of the skeleton!

Of course, everyone must try blowing through each of the skeletons and pulling the skeletons through the air. An optional challenge is to allow students to use a straw to blow bubbles within the geometric soap film shape.

What You Need for One Station

✓ 2 dish pans

✔ About 50 plastic coffee stirrers (either the one- or two-holed kind)

✔ 50–60 pipe cleaners

✔ 1 wire cutter or scissors for cutting pipe cleaners

✔ A hot glue gun

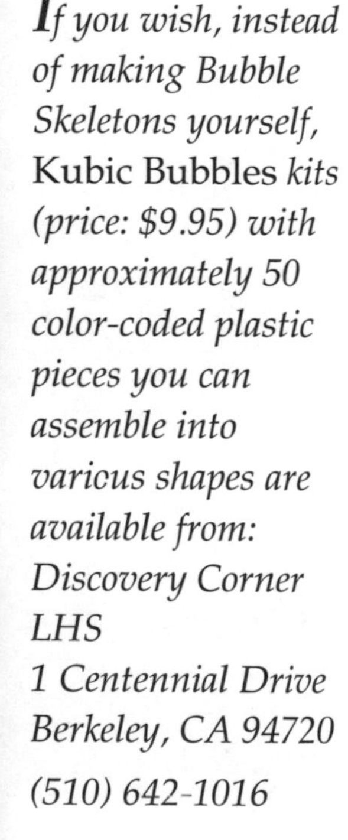
If you wish, instead of making Bubble Skeletons yourself, Kubic Bubbles *kits (price: $9.95) with approximately 50 color-coded plastic pieces you can assemble into various shapes are available from:*
Discovery Corner
LHS
1 Centennial Drive
Berkeley, CA 94720
(510) 642-1016

Getting Ready

☞ Make at least six skeletons per station, including, but not restricted to: two cubes, a pyramid (with three sides and a base), a six-sided shape made with two pyramids that share a base, and a pentagon.

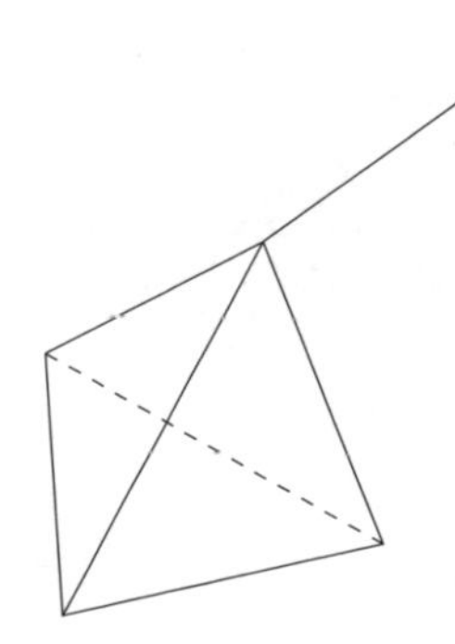

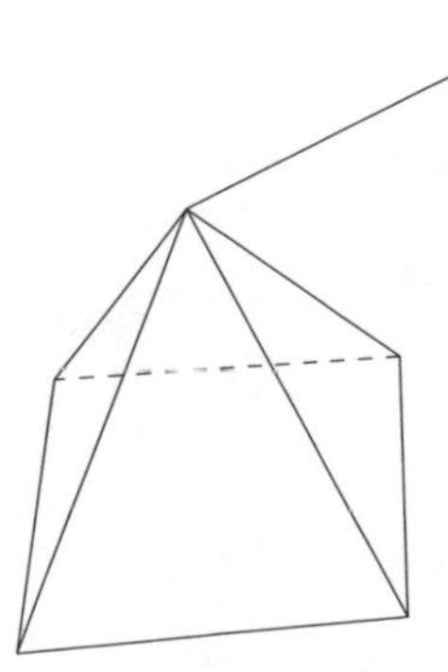

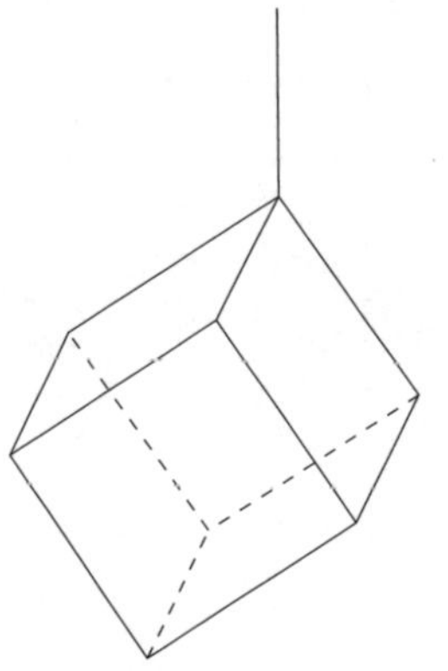

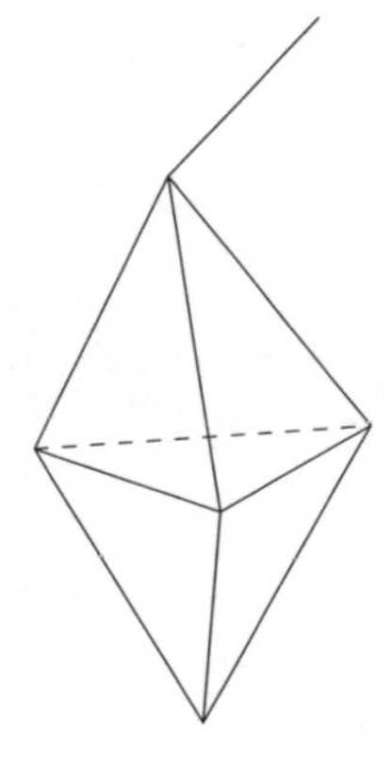

☞ To make the skeletons, cut the coffee stirrers in half. Cut a few pipe cleaners into segments that are about the same length as the half pieces of stirrers. Cut the rest of the pipe cleaners later, in case you decide that a slightly different length is better.

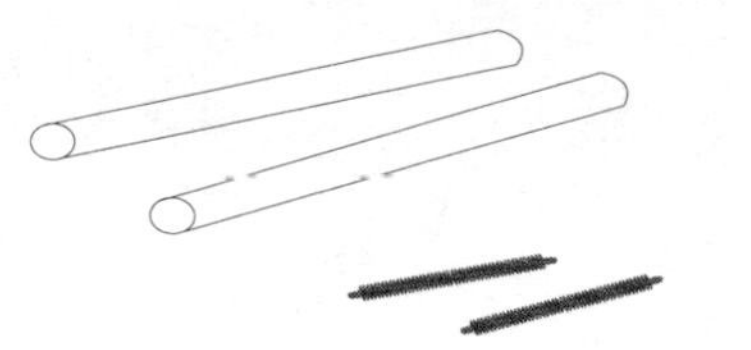

① Use one-half of a coffee stirrer segment per side. Connect two segments by inserting a pipe cleaner piece into both and then pushing them together; you can then bend the junction to make any angles you need.

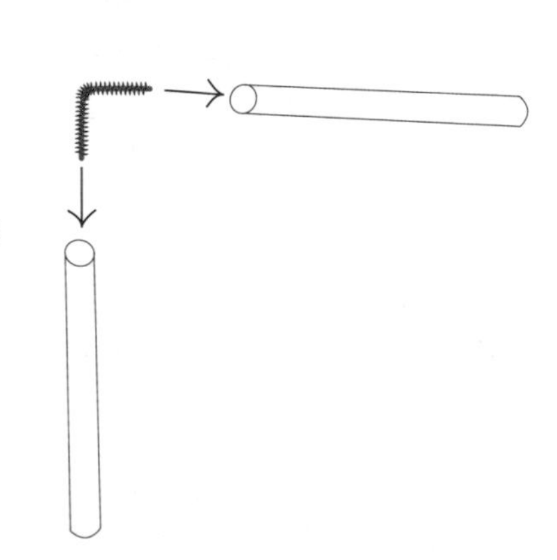

② Continue to connect the segments this way as you construct a polygon. In many cases, you will connect three or four segments at one corner by stuffing several pipe cleaner pieces into the stirrers. If a connection is wobbly, you may want to add an extra pipe cleaner piece. (This is more likely to be necessary with the one-holed kind of coffee stirrer than with the two-holed kind.)

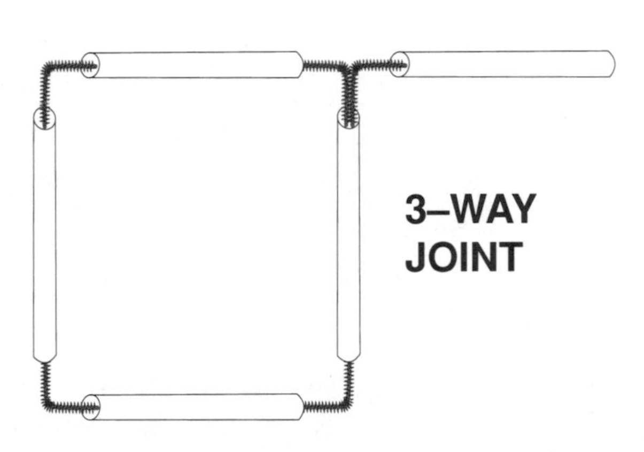

③ Use a hot glue gun to reinforce the joints and allow the bubble skeletons to dry.

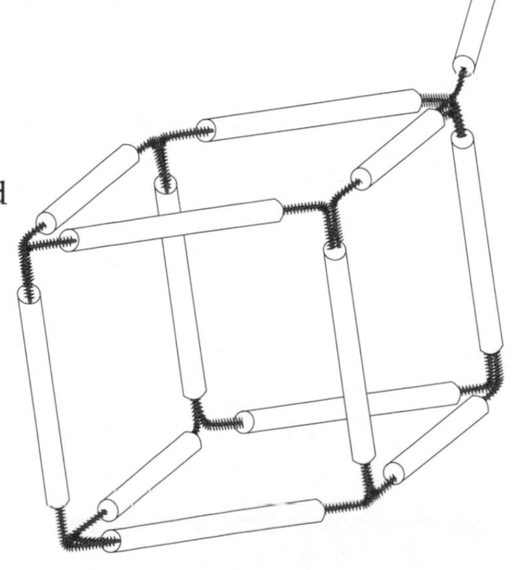

☞ Place two dish pans at the station and fill with enough bubble solution so that students will be able to completely immerse each skeleton.

☞ Put the skeletons at the station.

☞ Set out the "Bubble Skeletons" sign.

Special Considerations

- Make sure that the students understand these are not bubble wands to be waved about in the air. While they can be used as such, and students enjoy doing so, the most interesting phenomena occur as students dip the skeletons, observe the shapes that are made, and see how the shapes change as they use dry fingers to pop one face at a time. This activity gets messy when kids do wave the skeletons in the air. If possible, lay a drop cloth in front of this station and ask students to limit their waving to this area. Provide enough space between this and other activities so kids have room to move easily.

- Foamy soap solution interferes with this activity. You may want to ask your students not to jiggle the skeletons in the bubble solution as jiggling makes foam. Periodically skim the foam off the surface of the bubble solution.

Going Further

✌ Introduce your students to the names of the different polyhedrons that they experimented with. Your students may enjoy making coffee stirrer skeletons as a class activity.

✌ Conduct a demonstration with one of each polyhedron, and a bucket of bubble solution. Before you dip each shape into solution, ask the students to predict where the soap-film faces will adhere. Using a dry finger, pop the soap film one section at a time. Before you pop a surface, ask your students to predict if and how the rest of the soap films will change. As you do this with each of the polyhedrons, see if your students can predict what will happen and come up with a rule to explain why the soap film forms the shapes it does. With older students, introduce the idea of minimal surface area. See page 129 for more information.

✌ Have your students make tetrahedral kites or large polyhedrons to hang from the ceiling of the classroom.

Bubble Skeletons

What to Do

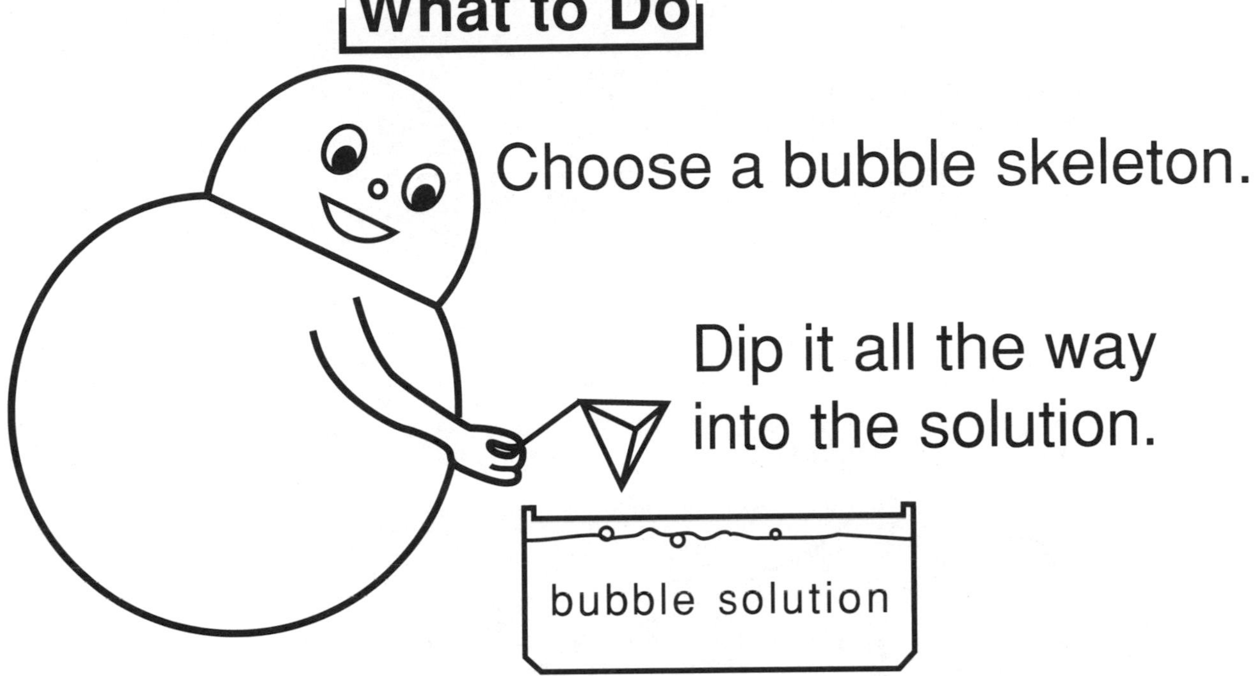

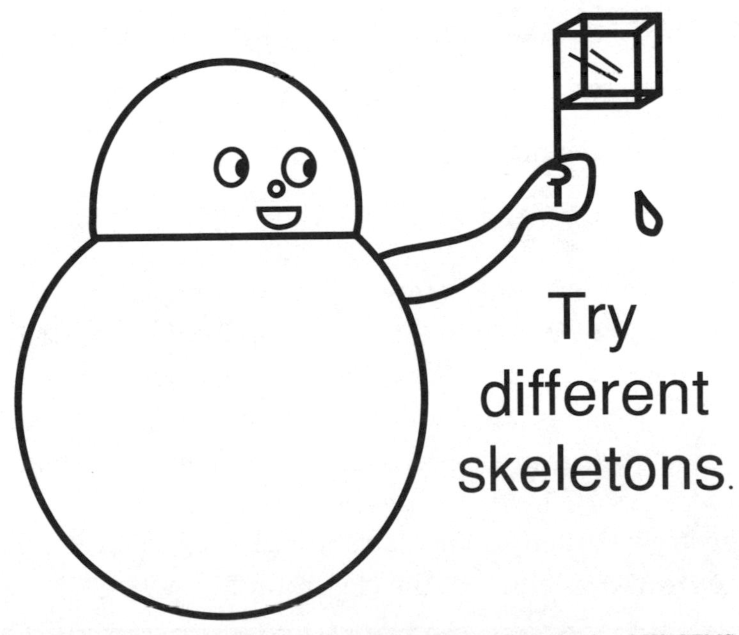

Bubble Skeletons

Questions

Try popping one side at a time.
How does the soap film change?

Can you make a little square bubble inside the cube skeleton? (Try double-dipping!)

Wave the skeletons through the air or blow into them with a straw.

What happens?

Activity Task Card for Volunteers

Bubble Skeletons

Your goal is to assist the students in making their own discoveries, while keeping the activity safe, and the mess under control. Read the signs at the station so you know what the students will be investigating. If the students are non-readers, you will have to communicate the content of each sign. This is best done by giving a challenge or asking a question, rather than demonstrating how it is done. Save your demonstrations for situations when students aren't successful on their own, even with coaching.

Ask the students open-ended questions, such as: What have you discovered? Why do you think that is happening? You may also want to provide further challenges, such as: Can you predict how the soap film will change when you pop one side? Why do you suppose it changes in that way?

Resist the temptation to give explanations to the students.

If students get out of control, involved in waving the skeletons too wildly or some other
activity that is unrelated to the station, you might want to steer them back on task. Make sure to intervene if you see an unsafe behavior. However, keep in mind that what may appear as "fooling around" with bubbles can lead to some of the deepest learning experiences. Some of the greatest scientific discoveries have been made while scientists "fooled around" — the same is true of great personal discoveries.

Tips for Managing the Station

- Remind students that all surfaces touching bubbles must be wet (hands, arms, etc.).
- Use a squeegee to periodically remove excess bubble solution from the table.
- Throw sections of newspaper over spills on the floor.
- Refill containers of bubble solution as needed. The solution in the dish pans should be high enough to cover the skeletons.

—Activity 10—

Frozen Bubbles

Overview

Students blow bubbles into a large, clear container with dry ice in the bottom. The bubbles will bounce and float in the container, then slowly grow in size! As the bubble gets bigger, it sinks in the container. Bubbles that land on the dry ice freeze instantly. People of all ages are fascinated by what is happening to these bubbles. Wondering why is at least half the fun! Don't spoil it by giving away "the explanation" to your students until they've had a chance to discuss their own ideas.

What is happening? The secret is in the dry ice. Dry ice is frozen carbon dioxide. As it warms, it changes from a solid to a gas, filling the container with CO_2. Since carbon dioxide gas is more dense than air, it accumulates first in the bottom of the container. Bubbles are a mixture of CO_2 and other gases from our lungs, so their contents are less dense than the CO_2 in the container. This is why the bubbles seem to float on this invisible "pillow" of gas.

Carbon dioxide is very soluble in water—this means that it easily passes through the soap film of a bubble wall. As the bubble sits in the container of CO_2, the CO_2 enters the bubble, causing it to grow bigger. Nitrogen and other gases in the bubble are much less soluble in water, so they remain in the bubble. As the bubble becomes filled with a greater concentration of CO_2, its density becomes closer to that of pure CO_2, thus the bubble sinks in the container.

Acknowledgment

This station is a modified version of an exhibit created by Ilan Chabay of the New Curiosity Shop in Mountain View, California.

What You Need for One Station

✓ 1 large, sturdy, clear plastic bucket, container, or plastic aquarium—cylindrical or square, about 4 gallon capacity
(Often restaurant supply stores will have clear plastic buckets, or you can call Rubbermaid at 1-800-347-9800 and ask for a distributor in your area that carries bucket model No. 6222-24)

✓ 1 slab of dry ice (Look under "Dry Ice" in the yellow pages)

✓ A cooler or stack of newspapers for insulation and transportation of the dry ice

✓ 1 pair thick, cotton gloves

✓ 1 or 2 cottage cheese containers

✓ straws

Getting Ready

☞ For success and safety, this station requires a full-time adult volunteer. Before the *Bubble Festival*, spend a few minutes with the volunteer going over the tips under "Special Considerations."

☞ After purchasing the dry ice, keep it in a cooler. If you do not have a cooler, wrap the dry ice in thick newspaper and fasten with tape or string. Always use gloves when handling dry ice. Direct contact with your skin can cause burns.

☞ About 15 minutes before the *Bubble Festival*, put the dry ice in the bottom of the clear, plastic container, and set it on the station.

☞ Just before the *Bubble Festival*, try blowing a bubble and letting it fall into the container to see if a layer of carbon dioxide gas has formed. (If the bubble doesn't float, take a stick and break up the dry ice a bit.)

☞ Fill the cottage cheese containers about half-full of bubble solution.

☞ Set out the "Frozen Bubbles" sign.

Special Considerations

- The adult volunteer will need to make sure that students keep their hands out of the container of dry ice; at 110° F below zero, it can burn their skin.

- Blowing downward into the container can disrupt the layer of carbon dioxide gas at the bottom. Instead, have the students take turns blowing bubbles above the container and observing them as they fall in.

- Ask questions such as the ones on the "Frozen Bubbles" sign to help students to describe and question what they observe. Avoid explaining "what's happening" until after the students have explored their own explanations.

- Stored under optimal conditions (in a 2"-3" thick-walled styrofoam container) dry ice can last up to three days. Wrapping dry ice in many layers of newspaper is also effective, but will keep the dry ice frozen for a shorter period of time.

Going Further

- Use thermometers to predict and measure the temperature of water, ice, warm water, and various mixtures of these.

For Upper Elementary

- Explore Bernoulli's principle and how air pressure affects bubbles in flight.

- Make carbon dioxide gas with vinegar and baking soda.

- Consider presenting activities from the GEMS guide *Discovering Density*.

Frozen Bubbles

What to Do

Blow a bubble over the bucket.
Watch as it falls into the bucket.

It works better
if you don't blow
down into the bucket.

NEVER TOUCH THE DRY ICE!

Frozen Bubbles

Questions

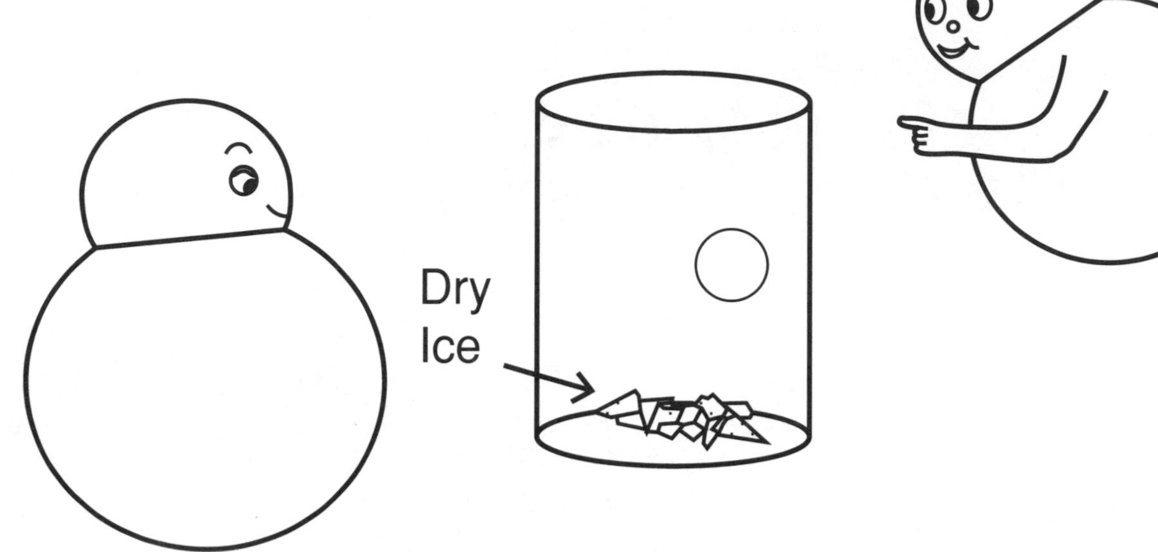

What happened to the bubble?
Where did it go?

Did the bubble change?
How?

Activity Task Card for Volunteers

Frozen Bubbles

Your goal is to assist the students in making their own discoveries, while keeping the activity safe, and the mess under control. Read the signs at the station so you know what the students will be investigating. If the students are non-readers, you will have to communicate the content of each sign. This is best done by giving a challenge or asking a question, rather than demonstrating how it is done. Save your demonstrations for situations when students aren't successful on their own, even with coaching.

Ask the students open-ended questions, such as: What have you discovered? Why do you think that is happening? You may also want to provide further challenges, such as: What is happening to the bubbles as time goes by? How are they changing?

Resist the temptation to give explanations to the students.

This is a station for quiet contemplation and discussion. Direct students to blow bubbles above the bucket and let them fall in. By blowing downward into the bucket, they may disrupt the layer of carbon dioxide gas that forms near the dry ice. **Be sure that no one puts their hands into the bucket.** Dry ice (frozen carbon dioxide) is so cold that it can "burn" the skin. Make sure to intervene if you see an unsafe behavior.

Tips for Managing the Station

- ✌ Use gloves when working with or near the dry ice.
- ✌ If a student blows into the bucket and disrupts the layer of carbon dioxide, just wait. The layer will re-form.
- ✌ Throw sections of newspaper over spills on the floor.
- ✌ Refill containers of bubble solution as needed.

Activity 11
Stacking Bubbles

Overview

What shape are bubbles when they are stacked together? Let your students blow bubbles into the narrow space between two clear sheets of plexiglass. How many sides do most of these crowded bubbles have? Do the walls intersect to make a "Y" shape?

What You Need for One Station

✓ 1 stacking apparatus, made from sheets of plexiglass clamped parallel to each with a stable base

✓ Rubber tubing (a diameter appropriate to receiving a drinking straw)

✓ A 2 foot by 3 foot tub to set the apparatus in.

Getting Ready

☞ Well before the *Bubble Festival*, construct the stacking apparatus. Put the rubber tubing into either side of the bottom of the plexiglass form, as shown in the drawing.

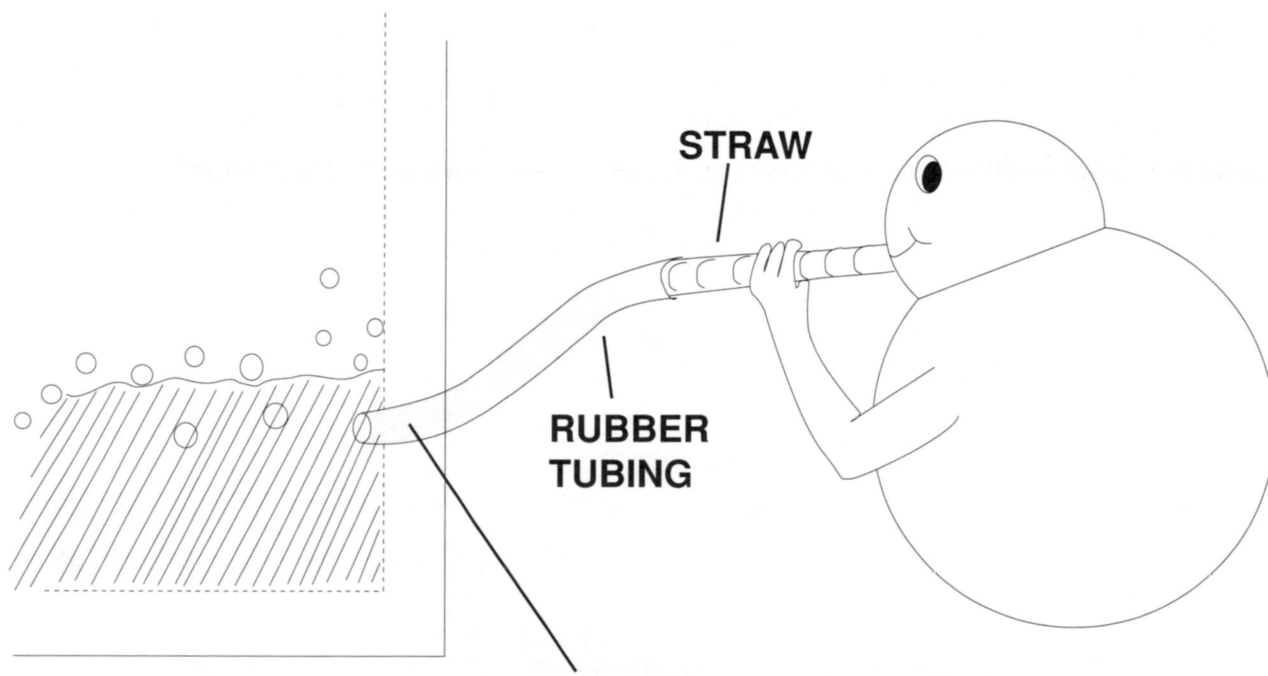

☞ Fill the tub with about two inches of bubble solution, and place the plexiglass structure into it.

☞ Set out the "Stacking Bubbles" sign.

Special Considerations

- Only two students at a time can blow into the bubble stacker; other students will watch until their turn comes. You may wish to restrict the number of students at this station to four.

- Students bring their own straws to insert into the rubber tubing when it's their turn to blow.

- Periodically remove the bubbles that have come out the top of the frame so that they don't obscure the view.

- Blow really softly!

Going Further

✌ Explore hexagons that are in nature and technology (honeycombs, airplane wings).

✌ Work with angles: Use a protractor to measure the degrees of each angle in a hexagon, and how they fit together. Which angles make strong, stable, lightweight structures?

Stacking Bubbles

What to Do

Put your straw into the rubber hose and blow.

Watch the bubbles in between the clear, plastic walls.

Stacking Bubbles

Questions

What shape are the bubbles when they are stacked?

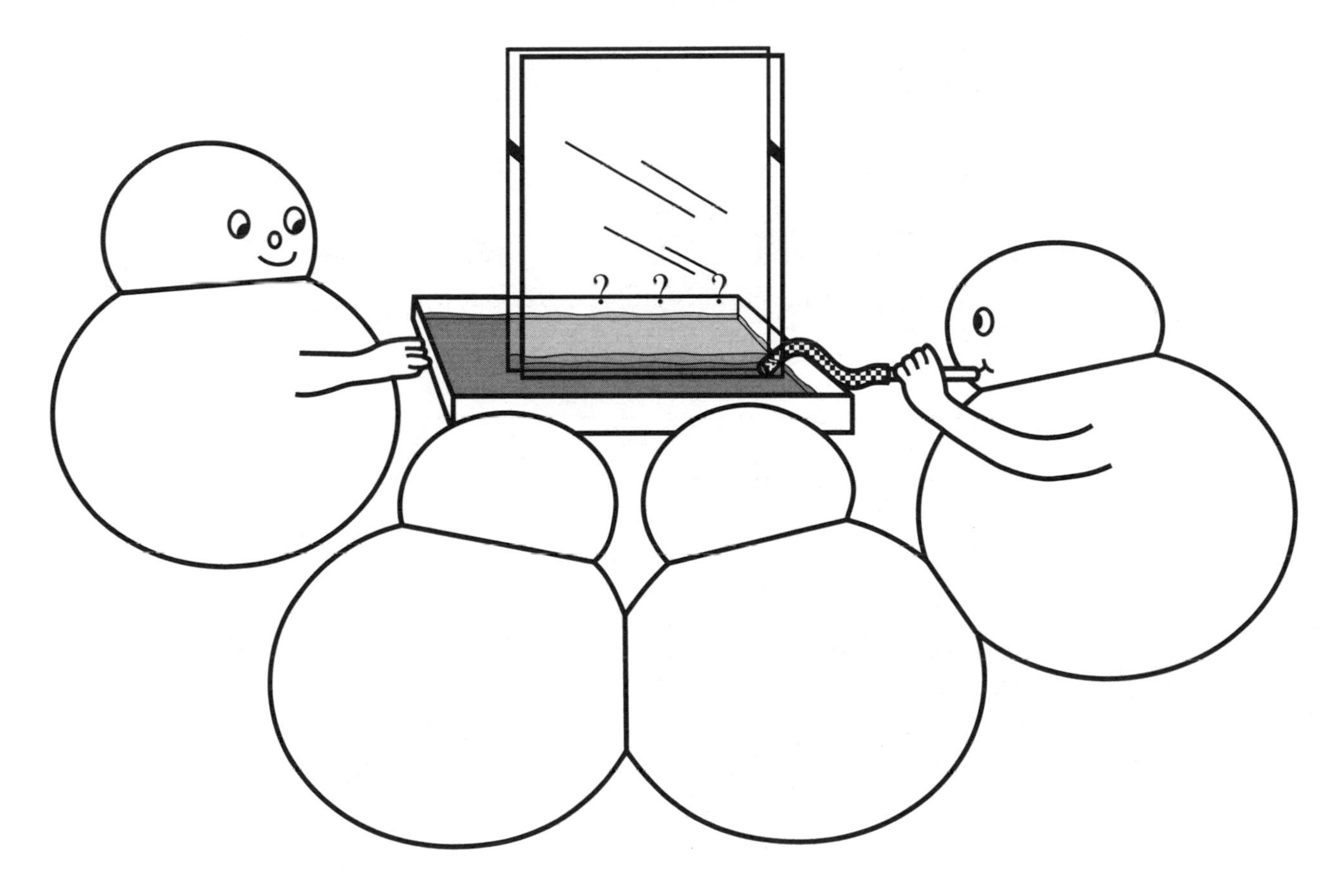

How many sides do most bubbles have?

Do the sides and corners of the bubbles join in any special way?

Activity Task Card for Volunteers

Stacking Bubbles

Your goal is to assist the students in making their own discoveries, while keeping the activity safe, and the mess under control. Read the signs at the station so you know what the students will be investigating. If the students are non-readers, you will have to communicate the content of each sign. This is best done by giving a challenge or asking a question, rather than demonstrating how it is done. Save your demonstrations for situations when students aren't successful on their own, even with coaching.

Ask the students open-ended questions, such as: What have you discovered? Why do you think that is happening? You may also want to provide further challenges, such as: Count the sides and "corners" of some of the bubbles. Do you see a pattern in the way the bubbles join together? Do the bubble patterns remind you of other things you've seen?

Resist the temptation to give explanations to the students.

If students get out of control, involved in creating mountains of foam or some other activity that is unrelated to the station, you might want to steer them back on task. Check to be sure that students are using their straws to blow, rather than blowing directly into the rubber tubing. Make sure to intervene if you see an unsafe behavior.

Tips for Managing the Station

- Check periodically to see that one end of the rubber tubing is placed between the sheets of plastic and **in** the bubble solution.
- Suggest that students blow very softly at this station. This will help them be more successful.
- Foamy soap solution interferes with this activity. Skim the foam off the surface of the bubble solution periodically.
- Refill the tub of bubble solution as needed. There should be enough bubble solution in the tub to cover the bottom edges of the sheets of clear plastic and the end of the rubber tubing.
- Use a squeegee to periodically remove excess bubble solution from the table.
- Throw sections of newspaper over spills on the floor.

—Activity 12—
Swimming Pool Bubbles

Overview

What is it like to be inside a bubble? Fun! At this station, a student stands on a platform in the middle of a wading pool filled with soap solution. The teacher or a parent volunteer raises a hula hoop to create a giant soap film cylinder around the child. Sometimes the cylinder seals at the top to form a complete bubble. Students of all ages are filled with wonder as they watch and wait their turn. How does the world look through a bubble? Try letting two students be "Bubble Buddies" inside one bubble, or connect two separate students with a bubble arch!

What You Need for One Station

✓ 1 plastic, "kiddie" wading pool about 3-4 feet in diameter

✓ 1 plastic hula hoop that will fit into the bottom of the pool

✓ 1 plastic milk crate or box (about 3 feet by 2 feet) that you can place in the middle of the pool full of solution for a child to stand on

✓ 1 roll duct tape

✓ 3 or 4 gallons of bubble solution (or whatever amount it takes to fill the pool about two inches high so that you can immerse the hula hoop.

✓ Plenty of newspapers or mats

Optional

1 old towel or T-shirt to cover the crate or box

Getting Ready

☞ Place the wading pool in an area that is away from drafts and where there is room for the students who will be standing in line for the activity.

☞ Using generous amounts of duct tape, tape the box or crate to the middle of the dry pool. Fill the pool with bubble solution about two inches deep. Put the hula hoop in the pool.

☞ Put out the "Swimming Pool Bubbles" sign.

Special Considerations

- This station requires a full-time adult volunteer to create the bubbles with the hula hoop and to assure that students get on and off the crate safely. With young children, two adults are preferable; one adult to make the bubbles and one to assist students in getting in and out of the pool.

- Sometime before the *Bubble Festival*, have the adult volunteer(s) practice enclosing a student in a bubble. The volunteer needs to practice using the following tips on safety and technique, and may also want to practice the **Going Further** ideas below.

Safety Tips

1) Before you begin, tell students to step on and off the crate *slowly*. Explain that they must not step off the crate until you tell them, because they may trip on the hula hoop. Assist students as they step on and off.

2) Keep the surface of the crate as dry as possible; if it is slippery, you may want to cover it with an old towel or T-shirt to provide traction.

3) If the floor around the pool is slippery, put down towels, stacks of newspaper or mats, as described in **On the Day of the Festival** on page 40.

4) To avoid back strain from repeated bending, try operating the hoop from a kneeling position.

Tips on Technique

1) Make sure the hula hoop and your hands are wet with bubble solution. Wet the insides of the pool and the sides of the crate, too.

2) Raise the hoop slowly. If the sides of the soap film cylinder pull inwards and touch the person inside, you are going too slowly.

3) Because the sides of the bubble do tend to pull inwards, have the student stand with hands down, holding in skirts or other bulky clothing. Try not to let feet protrude over the edges of the box or crate. See the sign for where and how to stand.

4) Periodically remove excess foam from the pool.

5) Some students have the impulse to close their eyes; suggest that they keep them open, because the bubble doesn't last long and they'll miss most of the fun of seeing what it's like to be inside a bubble.

Getting Fancy with Swimming Pool Bubbles

1) To close the bubble cylinder at the top, flip over the hoop high above the student's head (like flipping a pancake), and pull away. (This doesn't always work, and is not necessary for success.)

2) Try putting two "bubble buddies" in a bubble at once. Can you enclose more than two students?

3) Try enclosing the student and yourself in a giant bubble arch! After the upward motion of the hoop, make an arc that ends with the hoop going over your own head. You can also do this with another student who is standing nearby.

4) Train older students to raise the hula hoop out of the pool.

Going Further

- Use the **Going Further** suggestions found in "Bubble Windows" on page 86.
- Use the "Bubble Walls" learning station activities on page 92.

Swimming Pool Bubbles

What to Do

Get in line for
Swimming Pool Bubbles.

When it's your turn,
step carefully
onto the box.

Be sure to keep
your hands
down and your
eyes open!

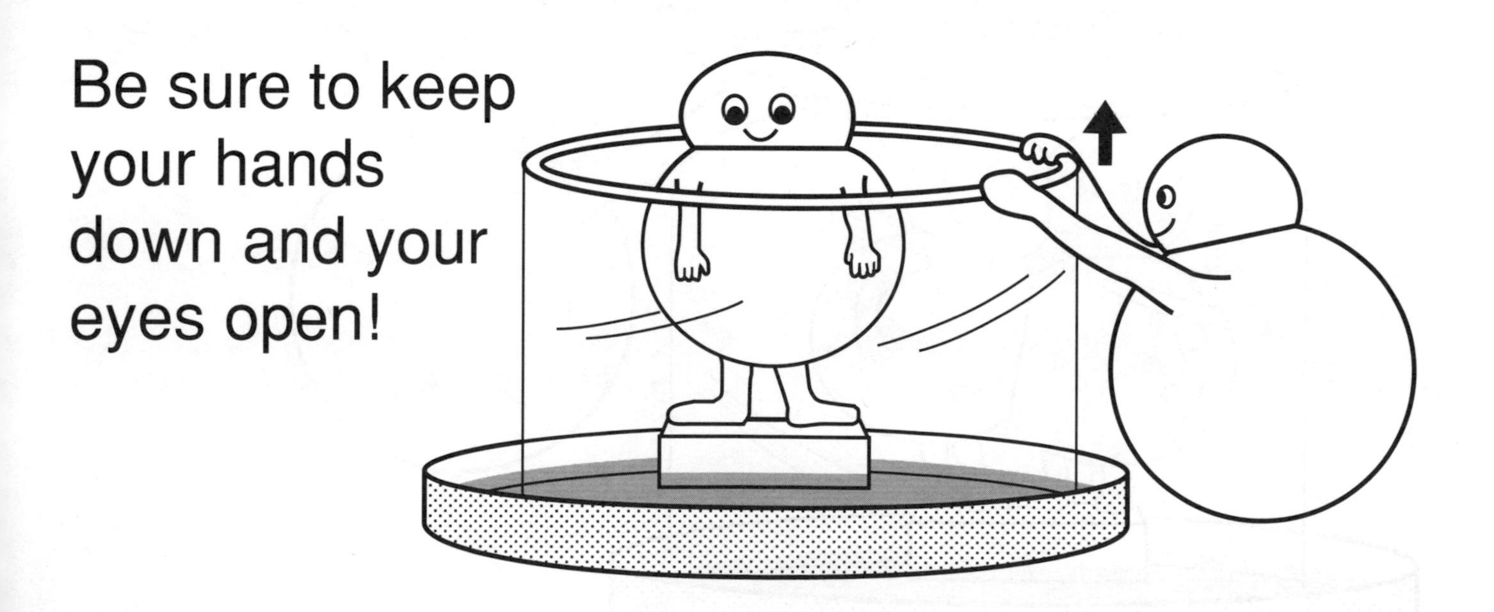

Swimming Pool Bubbles

Questions

What is it like to be inside a bubble?

Can you be a bubble buddy?

Activity Task Card for Volunteers

Swimming Pool Bubbles

Your goal is to assist the students in making their own discoveries, while keeping the activity safe, and the mess under control. Read the signs at the station so you know what the students will be investigating. If the students are non-readers, you will have to communicate the content of each sign. This is best done by giving a challenge or asking a question, rather than demonstrating how it is done. Save your demonstrations for situations when students aren't successful on their own, even with coaching.

Ask the students open-ended questions, such as: What is it like to be inside a bubble? What do you notice about the colors on the bubble film? You may also want to provide further challenges, such as: Why do you suppose the bubble film seems to pull inward?

Resist the temptation to give explanations to the students.

Students who are waiting their turn should stand in line and observe and discuss what they see. Help each student on and off the platform, reminding them to step carefully. Make sure to intervene if you see an unsafe behavior.

Tips for Managing the Station

- Before you begin, tell the students to step on and off the crate *slowly*.
- Keep the surface of the crate as dry as possible.
- Be sure that all surfaces that may touch the bubble are wet, including your hands and the inner sides of the wading pool.
- Remind students to hold in their arms and any bulky clothing.
- To avoid back strain as you bend to lift the hula hoop up around each child, you may want to kneel, rather than stand.
- Use a squeegee or a towel to remove excess bubble solution from the floor around the wading pool, and throw sections of newspaper over wet areas, especially where students mount and dismount from the pool.

Presenting *Bubble Festivals* to Several Hundred People

The Lawrence Hall of Science first presented our *Bubble Festival* to large groups, up to 150 people at a time. These were sometimes done for students during the school day, in the evening or on weekends for kids and their families, or at community events, including county fairs, end-of-sport season parties, and other celebrations. The festive feeling of these events generated much interest about the math and science that is embedded in most of the learning station activities.

The demand for these learning station activities was so great that we began presenting them to teachers and experimenting with how they could be used successfully in a classroom format. This guide is the result of that effort.

There are two different models for presenting a *Bubble Festival* to large numbers of people: the large-room model where a large group of people interacts in an unstructured way at a variety of learning stations, and the multiple-room model in which distinct activities are conducted in separate rooms, and smaller groups rotate through these rooms on a schedule.

Separate Rooms/Scheduled Rotation

By coordinating with other teachers, you can have a "*Bubble Festival* Day" at your school. Classes of students can rotate through each classroom on a schedule, while teachers stay put presenting the same activities each time. These can include learning stations involving bubbles or going further activities, some of which are writing or other "dry" assignments. While there is certainly effort in planning such an event, there is a great deal of time savings involved. Each teacher only has to prepare, present, and clean-up after one set of activities, and yet their students can experience six.

Large Room/Unstructured Interaction

The key to conducting a successful *Bubble Festival* for up to 150 people at a time is having enough stations, and an adult volunteer at every station. The Activity Task Cards for Volunteers, included in this guide, are invaluable in helping you organize 15 or 20 adults in a minimum of time. You'll be surprised how easy it is to get adult volunteers to come (a *Bubble Festival* is far more appealing than carpooling) and the adults will enjoy themselves as well.

How many stations are enough? That depends. If you have 150 kids coming, you need more stations than if you have 150 kids and parents. In the first case, figure that you will need enough stations with enough room at each station so every child can be at a station at all times. (For example, if you have tables with six stations, you would need 25 tables for a group of 150 kids.) With a mixture of parents and kids, there is more pairing up, and more watching and enjoying what each other is doing. If you want to contact 150 people, but don't have enough stations or a big enough room, consider having two 75-person sessions. Having even as many as four sessions is quite manageable. The main work is in the initial set up and final clean-up.

What other special arrangements are needed? Most of what's said in the earlier part of this guide applies to the large *Bubble Festival* as well. The main thing to consider is that, as a one-time event, many students don't get past the free exploration mode, except occasionally, when a particular station or challenge interests them enough to focus. And of course closure is difficult to provide, at least in the large group itself. There is not much you can do about either one of these limitations except to adjust your own expectations accordingly. There is more than enough learning and motivation taking place at a one-time *Bubble Festival* to make it worthwhile.

Following is a *Bubble Festival* morning conducted for six classes of 2nd–4th graders.

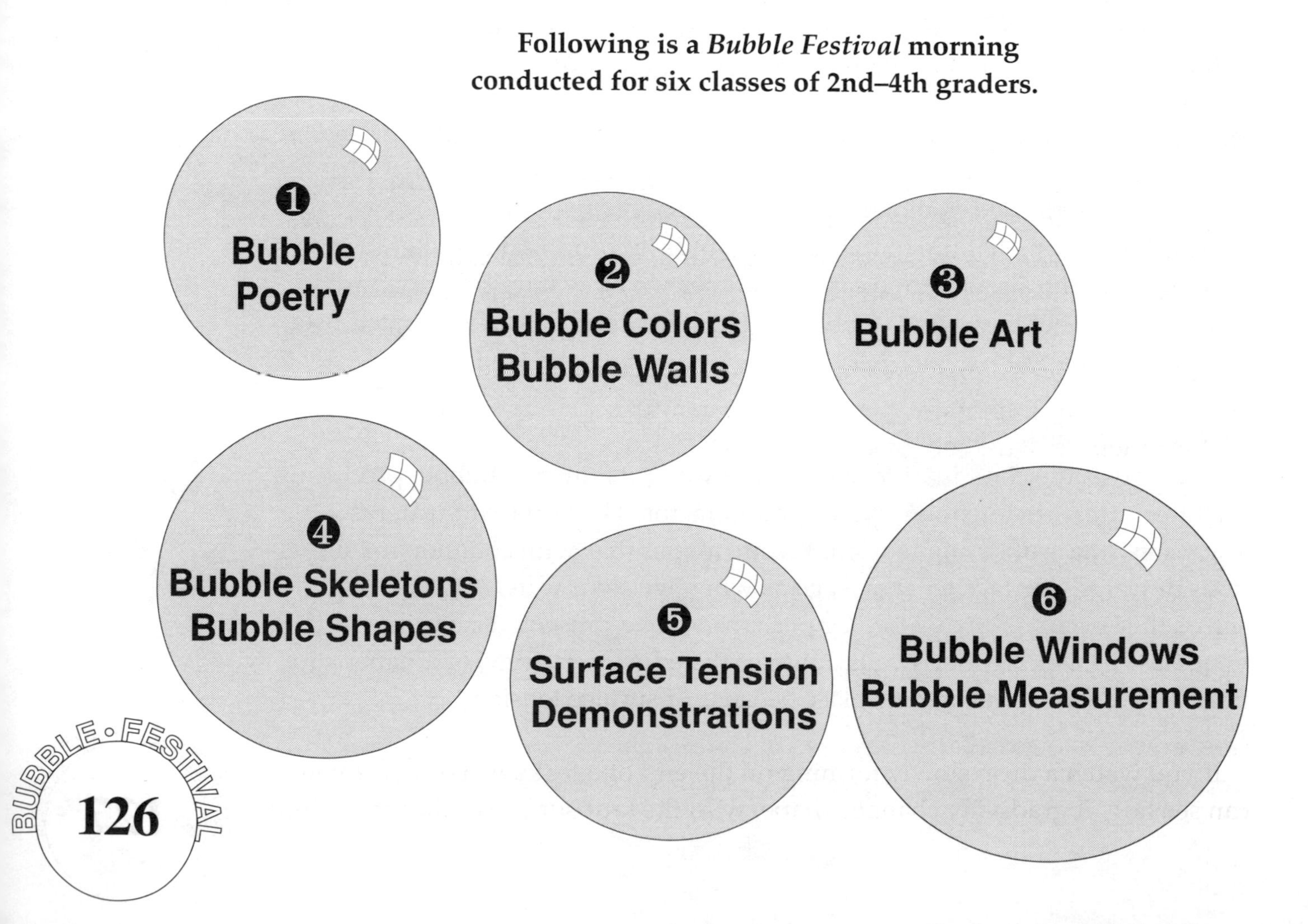

Behind the Bubble

Background to Bubbles

Bubbles and their behavior have fascinated people of all ages from time immemorial. Whether it be the foam upon the sea, the froth of root beer, or the first reflection of a soaring soap bubble in a baby's bright and curious eyes, bubbles are the "stuff" of poetry, of wonder and whimsy.

Yet bubbles are equally the "stuff" of science and mathematics. Their behavior has attracted and compelled observation by chemists, physicists, mathematicians, and engineers. The search for deeper understandings about bubbles has yielded intriguing connections in many technical fields; many books and articles have been written to consider these connections.

The resources listed at the end of this section can provide greater detail on specific subjects. This background is not meant to be read out loud to your students; it is provided to help you consider student questions and to give you a glimpse into this rich and diverse subject, as we skim along the shining surface to look into the scientific and mathematical marvels and forces that dance within and are reflected from—the bubble.

What is a Bubble?

A bubble is a thin skin of liquid surrounding a gas. This thin skin, or, in the case of soap bubbles, this **soap film**, has elastic qualities; it can stretch. The soap film is composed of molecules of water and soap. In the case of soap bubbles, the "gas" that the soap film surrounds is composed of either the gases that make up the air, or, in the case of bubbles we blow ourselves, the carbon dioxide and other gases that we exhale. The liquid and gas are separated by a **surface,** the soap film.

Surface Tension

What gives the combination of soap and water its unique, bubble-producing qualities? **Surface tension** is an important factor. The surface of water is always in a state of strong tension. We are all familiar with common manifestations of water's surface tension, such as when we see a water bug able to glide across the surface of the water. Or perhaps you've presented experiments to your class in which water swells out over the edge of a penny, or a paper clip floats. These phenomena take place because of surface tension.

If you watch a drop slowly forming on the end of a leaky water faucet you can see how it gradually changes shape, with the taut surface of the water drop at first containing the emerging water, then it begins to bulge out, then finally, when the surface tension of the water can no longer resist the downward pressure, a round drop detaches and falls. The forces of tension in the surface of the water at first contained the drop, but when it fell, the surface tension all around the drop then forced it into a sphere. (Another liquid, with less surface tension, would behave differently. For example, in Session 3 of the GEMS teacher's guide *Liquid Explorations*, students compare drops of oil and water, discovering that while water drops are round, oil drops are flat! Surface tension accounts for this difference.)

Surface tension is caused by the attraction a substance has to itself. Some molecules, such as water, have a positive electrical charge on one end and a negative charge on the other. These molecules can align, particularly on the surface of a liquid, so the positively charged end of one molecule forms an "attachment" with the negatively charged end of another molecule. These attachments are known as "**weak bonds.**" Their existence is what gives water its "stickiness" and what accounts for surface tension—at the surface, the bonds with other water molecules make the water behave as if it had a stretchy skin.

Along Comes Soap

When **soap** enters the scene, the two-sided nature of its molecules (with one side that is attracted to water and another side that is repelled by water and attracted to grease) creates important changes by *lessening* water's surface tension. In fact, soap reduces water's surface tension to about a third of what it usually is and some detergents cut the surface tension more than that. Here's how the reduction of surface tension happens:

(1) Soap molecules and water molecules are crowded together at the water's surface.
(2) The soap molecules near the surface have the water-attracting end next to the water molecules, but their water-repelling ends seek to push outward, into the air, as they are repelled by the water molecules.
(3) As these water-repelling ends of the soap molecules push outward, they also push between water molecules, spreading them apart. The wider apart these surface water molecules are, the less strong the hydrogen bonds between them. Thus, soap weakens water's surface tension.

The prevalence of the water-repelling (grease attracting) ends of the soap molecules near the surface also helps prevent the water from evaporating. This slowing down of **evaporation** is another factor that contributes to the ability of the soap film, the mixture of soap and water, to produce bubbles that last for some time.

Many bubble solutions also call for the addition of glycerin. Glycerin is a hygroscopic substance, meaning that it holds water, slowing evaporation. It thus intensifies the evaporation-slowing properties of soap. There is, however, an optimum amount, because *too much* glycerin can begin to have a counter-productive effect, and not contribute to making better bubbles. In the GEMS unit *Bubble-ology*, students experiment with differing amounts of glycerin to determine the optimum amount for making lasting bubbles.

Some people think that soap in combination with water helps make bubbles because it *increases* water's surface tension, but, as we have described, **the opposite is true.** In fact, two reasons why lasting bubbles cannot be made from water alone is that the surface tension of water is too strong and because evaporation takes place too fast. Soap and water together are loose and elastic enough to be stretched into a bubble, while water alone is not. **Soap reduces the water's surface tension and slows its evaporation, making it an ideal bubble-producer when mixed with water.**

Soap films also have the property of being self-healing. If a finger or other object is brought into contact with the soap film is also wet with soap solution, a moderately thick soap film will flow onto the wet finger or other object, thereby keeping the soap film continuous, and not popping the bubble.

Shapes and Surfaces

A bubble's **shape** is governed in part by the force of surface tension. The soap film enclosing a certain quantity of air stretches **only as far as it must** to balance the air pressure inside the bubble against the surface tension of the soap film. The bubble thus represents a **state of equilibrium** between air pressure and surface tension. The result is the spherical shape of bubbles. **The liquid forms the shape that encloses the greatest volume within the least surface area. This shape is a sphere**.

Soap films, such as the one the students experiment with in bubble windows and bubble walls, also provide excellent examples of the mathematical and physical concept of minimal surfaces. If you gently blow at or twist the surface of a film stretched across a "Bubble Wall" or "Bubble Window," it will stretch and its area will increase, but when you stop pushing, the soap film springs back to its original shape, taking up the smallest possible area that it can while still spanning the "window" or "wall" frame. Minimal surface area and surface tension are what cause the strings of the walls and windows to pull inward as well. Pulling on the "handles" of a bubble window can make a true square or rectangle. When left to hang, the soap film pulls the strings inward, resulting in a lower energy situation and a minimal surface arrangement.

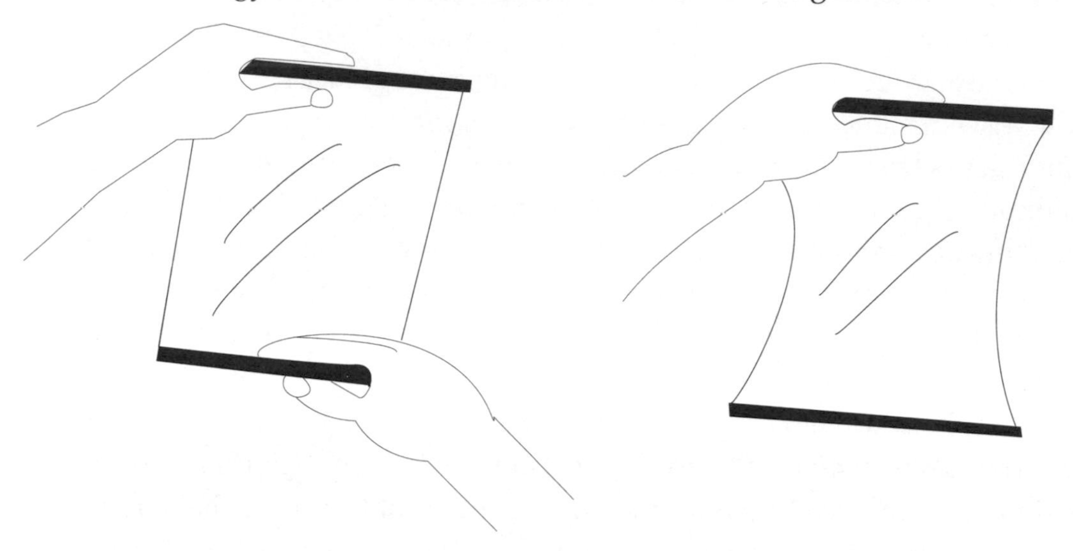

Students play with these interacting forces at the "Bubble Window" and "Bubble Wall" stations. Thinking about bubbles and other phenomena that behave in this way introduces topics like measurement of surface area, the relation of surface to area (which is always proportional in a bubble) and other physical science, mathematical, and geometrical concepts.

The edge of a soap film stretched across a wire frame is actually a solution to a complicated mathematical problem—finding the shortest possible path while obeying the laws of gravity and attaching to all surfaces dipped in soap. A very interesting example of this is provided when a wire frame in the shape of a cube is dipped in soapy water. The soap film sides that form can be forced to collapse in on the center, becoming a bulging cube that is untouched by the frame itself. The bubble tries to be round, but it cannot free itself from the frame. This has been described by Eiffel Plasterer as "a beautiful geometric compromise between a cube and a sphere." (Plasterer is perhaps the most famous "bubble-ologist" in the world; he holds the record for longest-living bubble—340 days—and has experimented with all kinds of bubble solutions and materials.) Others have compared a simple soap bubble to a computer, in the sense that soap film and bubbles instantly solve complex mathematical problems involving minimum paths and minimum surfaces with maximum volume, problems that would require a powerful computer to solve, and then not instantly. This is one of the reasons why so many scientific and technical disciplines have at one time or another used bubbles as models and simulations for physical or biological processes.

Polygons and Polyhedrons

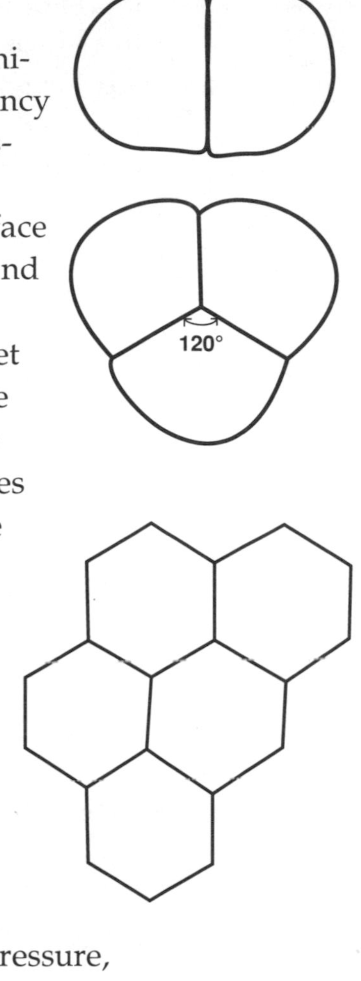

In the terms of modern physics, the soap film's capacity to seek minimal surfaces provides an excellent demonstration of a **system's** tendency to seek the state of lowest energy at which it can still function as a system. Soap bubbles are a system of interacting parts that include the liquid (soap and water), the gases enclosed, gravity, air pressure, surface tension, other properties, etc.). This minimizing of both surface area and systemic energy while maximizing volume also affects the ways that soap bubbles cluster together. When two bubbles meet, a smooth sheet separates them that is flat if the bubbles are the same size. When three bubbles come together, three surfaces intersect to form a line, and the angle between each pair of sheets is always 120°. Exploration of angles and their intersection, a key geometric concept, is another rich bubble math extension. The hexagonal-shaped bubbles formed by the intersection of bunched bubbles in the bubble wall are caused by this convergence of three sides. Interestingly, the hexagon is an efficient structure in nature, notably seen in the design of honeycomb. Students investigate these shape relationships in "Bubble Shapes."

But it doesn't stop there! The clustering of these hexagonal shapes, the stacking of bubbles, produces polyhedrons, or three-dimensional polygons. The German astronomer Johannes Kepler visualized the hexagonal crystalline pattern that comes from subjecting spheres to pressure, and he conceived of a twelve-sided figure with diamond-shaped faces like crystal garnet, called a rhombic dodecahedron. Compressed eggs and cells tend toward this shape; combining maximum volume with less partition area than almost any other multi-sided shape—nature at its most efficient. Lord Kelvin, the Scottish mathematician, then proposed a 14-sided figure, a tetrakaidekahedron (TKH) as even more ideal for living cells, with a slightly smaller percentage of partition surface. It took mathematicians a long time to conceptualize these "many-sided" ideas, but bubbles, cells, and crystals were already doing it! So, even in advanced realms of upper level mathematics, crystalline formation, and cell biology, scientists learn and continue to explore a lot from the behavior of bubbles!

Topology and Architecture

The special properties of soap film, including numerous experiments with intricate loops and shapes and soap film, brings us to topology. Topology is the exploration of the geometrical properties of various shapes, substances, and figures, especially relating to what happens when shapes, substances, and figures are bent, stretched, or molded. This fascinating subject is explored in experiments such as the "Bubble Skeletons" activity in this guide.

Using soap films and soap-film computer simulations, mathematicians have found complex minimal surface forms called catenoids and helicoids, and their work has already found adaptations in dentistry, embryology, and architecture. For example, architect Frei Otto, who designed structures at the Munich Olympics, utilized soap film models to design graceful, airy, and web-like structures, using as little construction material as safely possible to create exhibition halls, arenas, and stadiums that could be easily built, dismantled, and moved. Otto also written a book on tensile structures that we list below.

Light and Color

The beautiful color patterns seen when a beam of light shines through a soap bubble are produced whenever light passes through extremely thin layers. This is similar to an oil slick on the street, or even a peacock's tail feathers. The colors come from the reflection of white light shining on the bubble. White light contains waves of different colors—the spectrum. The length of a wave determines its color.

When light bounces off the bubble, some of each wave is reflected from the outer surface of the bubble wall, and some passes through the bubble and is reflected by the inner surface. This means there are two sets of light waves passing through the same space at the same time, and causes a phenomenon called **interference**. Sometimes the crests of the two sets of waves coincide, intensifying both (**constructive interference**); sometimes the crest of one meets the trough of another and that particular color is cancelled out (**destructive interference**). As the bubble wall thins, the wavelengths that are seen, or not seen, also change.

Just before it bursts, the bubble appears white with a growing black (or transparent) spot. This is because, when the wall is less than a quarter wavelength thick, none of the colors is completely cancelled out, all are present, so the bubble appears white. A black spot or dot appears when the bubble is super-thin (about one-millionth of an inch) because, due to the (destructive) interference phenomenon at that point, all the waves from the front surface and back surface cancel each other out. The black dot gradually expands, then disappears as the bubble pops.

There are many experiments that can be done exploring the fascinating and colorful interactions between light and color. In this guide, students make unique observations in the "Bubble Colors" activity. In the GEMS teacher's guide *Bubble-ology* students "Predict-a-Pop" in Session 5, gaining practical experience and getting a great introduction to important concepts in physics, while exploring the same wondrous phenomena represented by the rainbow.

A Vast Spectrum of Real-Life Applications

In the standard tool chest, we find many types of levels that utilize bubbles. You could bring one or more of these into class. Bubbles are also used in some sextants for navigation and surveying.

Bubble shapes, minimal energy and surface qualities, the formation of polyhedrons, and other bubble behavior has found application in many engineering and architectural fields, including the construction of tents, geodesic domes, larger structures, and as models for the solution of a wide range of engineering problems. A good example is designing a tin soup can—what is the best shape and minimum amount of material that can be used in packaging? Bubbles have helped engineers figure this out.

Even bubble noise, the sound of tiny air bubbles bursting, has been explored as a way of detecting submarines. It is heard by scuba divers as a high-pitched tinkling sound. Liquid drop models, one form of bubbles, have been used in mathematical models studying the nucleus of the atom.

The study of bubble formation and behavior, including "soap froth," has also provided useful models for further understanding the shape of developing embryos, the chemical structure of polymers, physical processes in crystallography, and the structure of metals. In dentistry, it has been suggested that least-surface area shapes could be used to design bone implants for securing false teeth, to minimize contact with the bone while maximizing a strong bond.

We are all familiar with the rushing bubbles in boiling water. In nuclear physics, scientists use the phenomena of boiling bubbles in a bubble chamber to see the pathways of invisible atomic particles. The paths left by high-speed protons going through liquid hydrogen as the pressure is reduced just enough for the very first boiling to occur are tiny trails of bubbles. The invention of the bubble chamber by Donald Glaser (who received the Nobel Prize in 1955) has helped make possible many new developments in particle physics and further widened our understanding of atomic structure.

But there is no need to look far afield for bubbles. From the bubbles in carbonated drinks, beer, or sparkling champagne, to the yeast bubbles that form in the baking of bread, bubbles are part of our everyday life. Hot air balloons and blimps were the first ways people invented to fly. Air bubbling into an aquarium partly dissolves into the water and enables the fish to breathe. A light bulb is a bubble of inert gas that keeps oxygen away from a glowing filament, preventing it from burning up quickly. Firemen use foam to put out fires, blocking off oxygen, and releasing cooling moisture as the bubbles pop.

Plastic bubble "pak" is used for mailing fragile items (and is fun to play with besides!).

Last, but perhaps not least, it has been suggested by several scientists, including astrophysicist Margaret Geller, that some astronomical computer tracking and computation indicates that the entire universe itself may be composed of bubbles. These theories speculate that the planets, stars, and galaxies we know are actually on the surface of these giant bubbles. It is even possible that these bubbles may at times behave like soap bubbles, sometimes merging into each other!

Today,
***Bubble Festival* learning station activities**
in the classroom,
tomorrow the universe!

Writing and Bubbles

Bubbles provide a fabulous context for writing. There are spectacular physical phenomena to describe that evoke both multi-sensory descriptions and fantasy accounts. There are observations, procedures, experiments, and discoveries to share and explain. The writing can take many forms: word banks, phrase charts, sentence completions, quick-writes, open-ended journal writing, poems, stories, responses to open-ended questions, or lab write-ups. The sky's the limit. Teachers comment again and again that their students can't write enough about their experiences with bubbles. Following is a collection of ways various teachers have used writing in conjunction with their *Bubble Festivals*.

Create a Chart of Bubble Experiences

—I can make round bubbles by blowing through square holes.
—I can make long bubbles open on end.
—When you pop a bubble in the air it sometimes leaves a wispy skin.

Sentence Starters

—I was amazed when ...
—The best part of Body Bubbles was ...
—One thing I learned about bubbles was ...

Have Students Write:

—In their journals about the question "Why do bubbles always come out round?"
—Step-by-step instructions for how to blow the perfect table bubble. (Find a buddy to test those instructions!)
—A story about a world that had bubble foam instead of grass.
—A story about life without any type of bubbles. What would we not have? (bread, carbonated drinks, foam rubber, shaving cream, etc.)
—A letter instructing someone on how they can use color to learn about a bubble. (This is for older students.)
—A fantasy. "I looked through my bubbly, hazy, window, and I saw ..." or fantasize about what's beyond the bubble wall.
—A list all of the words you can think of to describe the bubble shapes you saw.

Have Students Write Poems:

—Entitled "Me in My Bubble" after doing the activity "Swimming Pool Bubbles" in which each student is encased in a giant bubble.

Write a Bubble Biography:

—About the story of how a bubble is born, its travels, and finally, how it died. How did its colors change? Its shape?

Write a bubble dictionary:

—All the words a bubble-ologist might use and definitions for those words.

Dictate stories:

—About what they saw (This is for younger students.)

Publish a bubble book:

—A collection of class discoveries, new questions, and pictures of bubbles.

Children's Literature Connections to the *BUBBLE FESTIVAL*

MONSTER BUBBLES
by Dennis Nolan
New Jersey.: Prentice-Hall, Inc. (1976).
(Grades: Preschool–1)

Various monsters take turns blowing bubbles in amounts from one to twenty. A fun counting book for younger students.

BUBBLE BUBBLE
by Mercer Mayer
New York: Parent's Magazine Press (1973).
(Grades: Preschool–2)

A little boy buys a magic bubble maker, then creates and controls bubbles in the shapes of progressively larger animals.

THE MAGIC BUBBLE TRIP
by Ingrid & Dieter Schubert.
New York: Kane/Miller Book Publishers (1981).
(Grades: Preschool–3)

James blows a giant bubble that carries him away to a land of large hairy frogs where he has a fun adventure.

HAILSTONES AND HALIBUT BONES: ADVENTURES IN COLOR
by Mary O'Neill; illustrated by John Wallner
New York: Doubleday (1961)
(All grades)

Twelve poems about various colors, painting a wide spectrum of images. Good connection to the observations of color changes in bubbles in this guide (and to the "Predict-a-Pop" activity in the GEMS unit *Bubble-ology*).

MR. ARCHIMEDES' BATH
by Pamela Allen
New York: Lothrop, Lee & Shepard.(1980).
(Grades: K–2)

Upset by his bath overflowing and puzzled by the changing water level, Mr. Archimedes first tries to blame one of his three bath companions (a kangaroo, a goat, and a wombat). Then he resorts to scientific testing and measuring to find out. Good connections to measurement activities, and to questions about volume and density.

HOW BIG IS A FOOT?
by Rolf Myller
New York: Dell/Bantam (1990)
(Grades: K–5)

Delightful story about an apprentice carpenter and the need for a standard unit of measurement when building a bed for the queen. A good connection to the bubble measurement activities in this guide.

SPLASH! ALL ABOUT BATHS
by Susan Kovacs Buxbaum and Rita Golden Gelman; Illustrated by Maryann Cocca-Leffler
Boston: Little, Brown, and Co. (1987).
(Grades: K–6)

Before he bathes, Penguin answers his animal friends' questions about baths, such as: "What shape is water?" "Why do soap and water make you clean?" "What is a bubble?" "Why does the water go up when you get in?" "Why do some things float and others sink?"

SCIENCE ON STAGE ANTHOLOGY: THE SOAP OPERA
by LHS staff
Washington, D.C: Association of Science-Technology Centers (1991)
(Grades: 3–7)

This play is an exciting molecular romp through the chemistry of soap and how soap cleans. It would be excellent as an entertaining and educational finale for your *Bubble Festival*.

RUBBER BANDS, BASEBALLS, AND DOUGHNUTS: A BOOK ABOUT TOPOLOGY
by Robert Froman; illustrated by Harvey Weiss
New York: Crowell (1972.)
(Grades: 4–10)

Concise discussions of topological topics that can be related to the behavior of soap film in the bubble skeletons, walls, and windows activities.

More About Bubbles: A Bubble Bibliography

Almgran, Frederick J. Jr., and Taylor, Jean E. "The Geometry of Soap Films and Soap Bubbles." *Scientific American*. (July 1976) 235: 82.

"Bubbles, Foam, and Fizz." *Ideas in Science*. (January 1983).

"Geometry in Nature: Bubbles/Mathematics in Nature." *Ideas in Science*. (Vol. 3, No. 2, 1986).

Bohren, Craig F. *Clouds in a Glass of Beer: Simple Experiments in Atmospheric Physics*. New York: John Wiley & Sons, 1987.

Boys, C.V. *Soap Bubbles: Their Colors and Forces Which Mold Them*. New York: Dover, 1959.

Chemistry with Bubbles. Mountain View, California: The New Curiosity Shop, Inc., 1988. (In cooperation with Apple Computer, Inc.).

Clift, R. Grace, J.R., and Weber M.E. *Bubbles, Drops, and Particles*. San Diego: Academic Press, 1978.

"Bubbles." *The Exploratorium Magazine*. (Winter 1982).
(Copies available from The Exploratorium, 3601 Lyon Street, San Francisco, CA 94123.)

Isenberg, Cyril. *The Science of Soap Films and Soap Bubbles*. New York: Dover, 1978.

Katz, David A. *Chemistry in the Toy Store*. Department of Chemistry, Community College of Philadelphia. Second edition, 1983. (Copies available from the author.)

Levine, S., Strauss, M.J., Mortier, S. "Soap Bubbles and Logic." *Science and Children*. (May 1986) 23:10-12.

Murphy, Jamie. "Bubbles in the Universe." *Time*. (January 20, 1986): 51.

Otto, Frei. *Tensile Structures*. Cambridge, Massachusetts: Massachusetts Institute of Technology Press, 1972.

Peterson, Ivars. "Science Meets the Soap Bubble." *Washington Post*. (September 18, 1988): C3.

Rice, K. "Soap Films and Bubbles." *Ideas in Science*. (May 1986): 23:4–9.

Siddeons, Colin. "Soap Bubble Spectra." *The Science Teacher*. (January 1984): 26.

Steinhaus. H. *Mathematical Snapshots*. New York: Oxford University Press, 1983.

Stevens. Peter S. *Patterns in Nature*. Boston: Little, Brown and Company, 1974.

Zubrowski, Bernie. Illustrated by Joan Drescher. *Bubbles: A Children's Museum Activity Book*. Boston: Little, Brown and Company, 1979.

Zubrowski, Bernie. "Memoirs of a Bubble Blower," *Technology Review*. (November/ December 1982).

Body Bubbles

What to Do

Try using your fingers to blow a bubble in the air …

… or on the table.

Try using two hands to make bubbles.

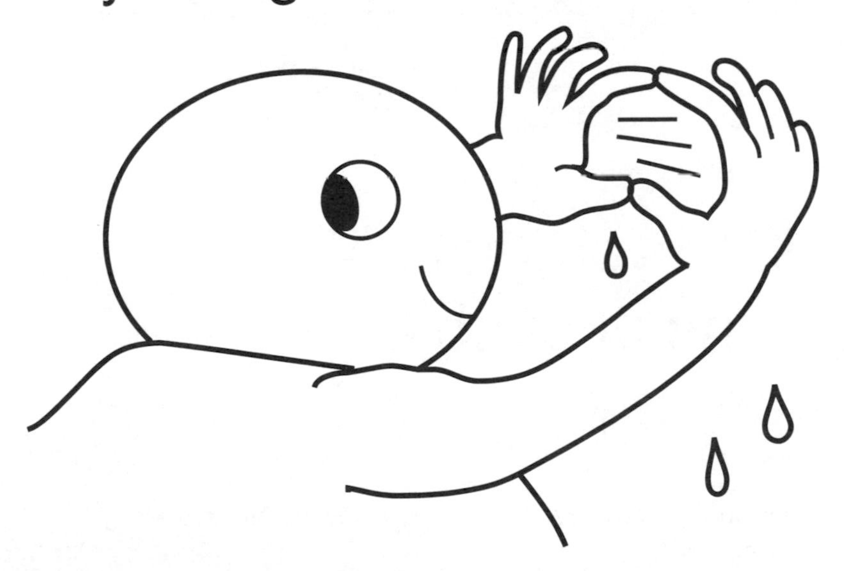

Use a straw to blow bubbles in your hand!

Body Bubbles

Questions

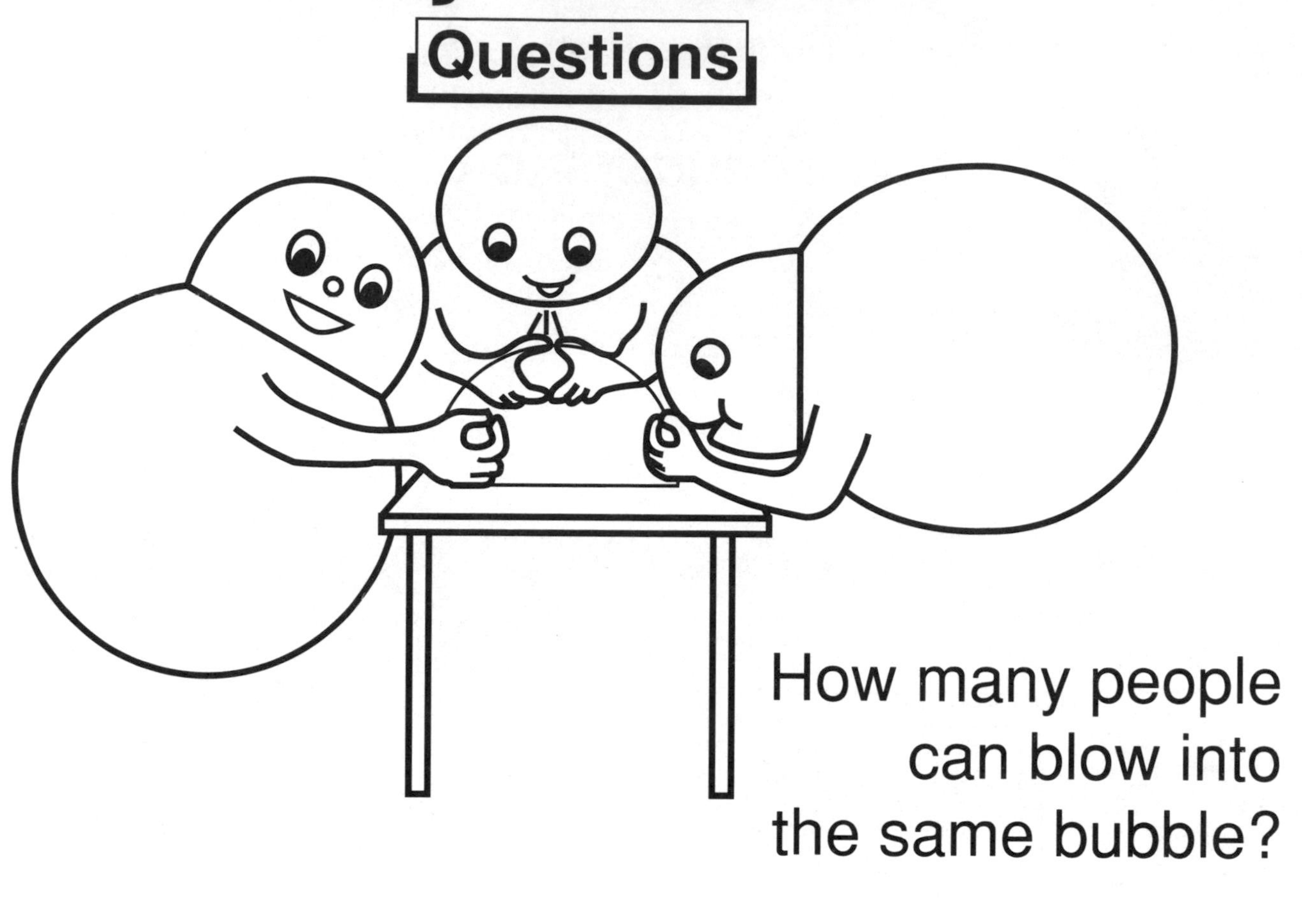

How many people can blow into the same bubble?

Can you make a line of bubbles down your arm?

Activity Task Card for Volunteers

Body Bubbles

Your goal is to assist the students in making their own discoveries, while keeping the activity safe, and the mess under control. Read the signs at the station so you know what the students will be investigating. If the students are non-readers, you will have to communicate the content of each sign. This is best done by giving a challenge or asking a question, rather than demonstrating how it is done. Save your demonstrations for situations when students aren't successful on their own, even with coaching.

Ask the students open-ended questions, such as: What have you discovered? Why do you think that is happening? You may also want to provide further challenges, such as: Can you think of a different way to use your hands to make a bubble? Can you use four hands to make a bubble? Six hands?

Resist the temptation to give explanations to the students.

If students get out of control, involved in creating mountains of foam or some other activity that is unrelated to the station, you might want to steer them back on task. Make sure to intervene if you see an unsafe behavior. However, keep in mind that what may appear as fooling around with bubbles can lead to some of the deepest learning experiences. Some of the greatest scientific discoveries have been made while scientists "fooled around"—the same is true of great personal discoveries.

Tips for Managing the Station

- Remind students that all surfaces touching bubbles must be wet (hands, arms, etc.).

- Use a squeegee to remove excess bubble solution from the table surface periodically.

- Throw sections of newspaper over spills on the floor.

- Refill containers of bubble solution as needed.

Bubble Shapes

What to Do

Blow a bunch of bubbles that touch ...

... on the table ...

... or in your hand.

Look at the places where bubbles meet.

Bubble Shapes

Questions For Younger Students

Can you make a bubble with straight sides?

Can you make a bubble that looks like a box?

What shapes do you see?

Can you blow a bubble inside a bubble ... inside a bubble ... inside a ...

Bubble Shapes

Questions

Can you make a bubble with four sides?
Five?

Can you make a
six-sided cube
out of bubbles?

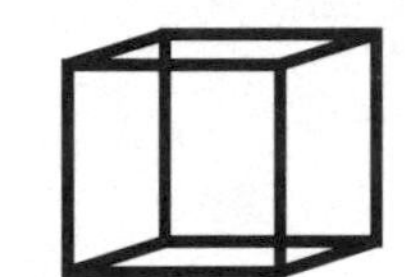

What angles
do you see?

Can you blow a bubble
inside a bubble …
inside a bubble …
inside a …

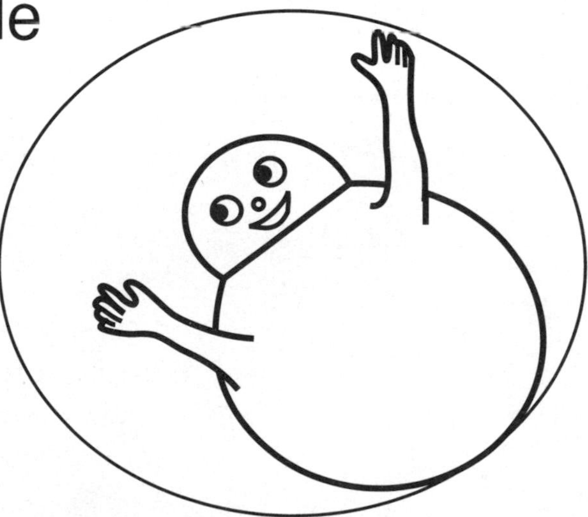

Activity Task Card for Volunteers

Bubble Shapes

Your goal is to assist the students in making their own discoveries, while keeping the activity safe, and the mess under control. Read the signs at the station so you know what the students will be investigating. If the students are non-readers, you will have to communicate the content of each sign. This is best done by giving a challenge or asking a question, rather than demonstrating how it is done. Save your demonstrations for situations when students aren't successful on their own, even with coaching.

Ask the students open-ended questions, such as: What have you discovered? Why do you think that is happening? You may also want to provide further challenges, such as: What's happening to the bubble as you squeeze it? Can you make a triangular bubble?

Resist the temptation to give explanations to the students.

If students get out of control, involved in creating mountains of foam or some other activity that is unrelated to the station, you might want to steer them back on task. Make sure to intervene if you see an unsafe behavior. However, keep in mind that what may appear as fooling around with bubbles can lead to some of the deepest learning experiences. Some of the greatest scientific discoveries have been made while scientists "fooled around"—the same is true of great personal discoveries.

Tips for Managing the Station

- Remind students that all surfaces touching bubbles must be wet (hands, straw, table), to dip the straw in solution just prior to blowing, to blow very softly, and not to inhale with their mouths on the straw.

- If you have a student who has difficulty blowing a table bubble, try blowing a bubble into his soapy hand so he can feel just how softly he needs to blow. Suggest he begin by blowing a bubble into his own palm prior to blowing a table bubble. If a student takes a quick breath in with her mouth on the straw, she is breaking the soap film at the end of the straw. Cut a diamond-shaped hole in the straw an inch or so below the blowing end to eliminate the problem.

- Keep the table surface covered with a thin coating of bubble solution. Use a squeegee to periodically remove excess bubble solution from the table.

- Throw sections of newspaper over spills on the floor.

- Refill containers of bubble solution as needed.

Bubble Measurement

What to Do

Blow a table bubble. When it pops, it will leave a circle on the table—a bubble print!

The distance across the widest part of the bubble print is called the diameter.

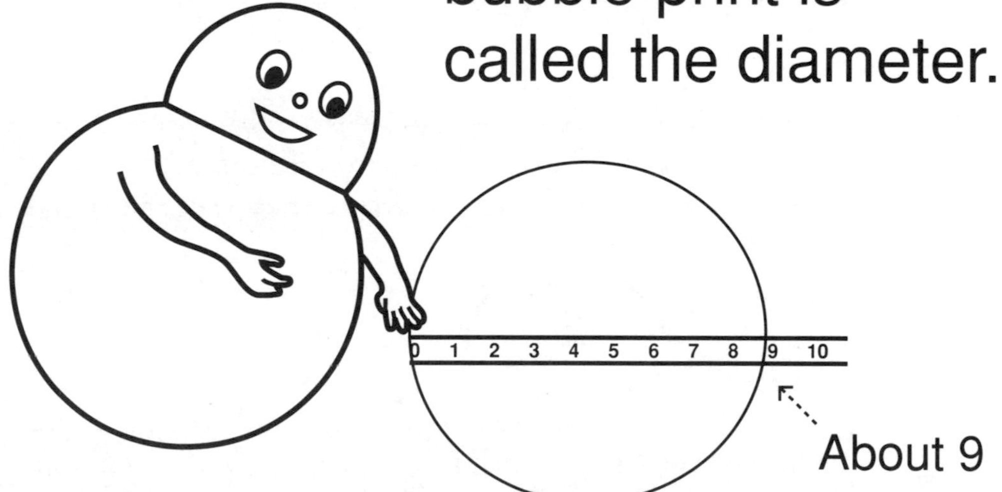

Try measuring the diameter of your bubble with some of the measuring tools at the station, or even with your hand.

Bubble Measurement

What to Do For Younger Students

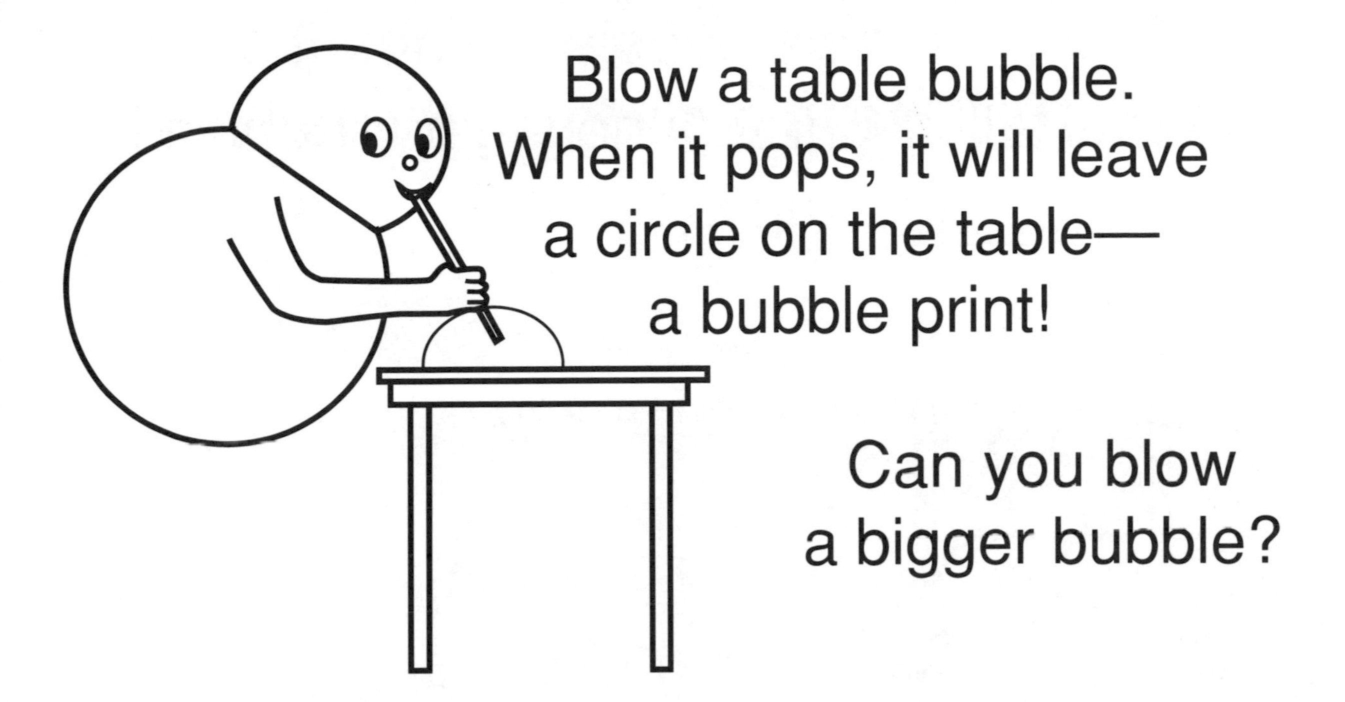

Blow a table bubble.
When it pops, it will leave
a circle on the table—
a bubble print!

Can you blow
a bigger bubble?

How many bubbles will it take
to cover the table?

Bubble Measurement

Questions

How tall or big around is your bubble?
How big is the space inside it?

Try three different ways
of measuring your bubble.

Activity Task Card for Volunteers

Bubble Measurement

Your goal is to assist the students in making their own discoveries, while keeping the activity safe, and the mess under control. Read the signs at the station so you know what the students will be investigating. If the students are non-readers, you will have to communicate the content of each sign. This is best done by giving a challenge or asking a question, rather than demonstrating how it is done. Save your demonstrations for situations when students aren't successful on their own, even with coaching.

Ask the students open-ended questions, such as: What have you discovered? Why do you think that is happening? You may also want to provide further challenges, such as: Can you think of another way to measure a bubble? What's the biggest bubble you can make? The smallest?

Resist the temptation to give explanations to the students.

If students get out of control, involved in creating mountains of foam or some other activity that is unrelated to the station, you might want to steer them back on task. Make sure to intervene if you see an unsafe behavior. However, keep in mind that what may appear as fooling around with bubbles can lead to some of the deepest learning experiences. Some of the greatest scientific discoveries have been made while scientists "fooled around"—the same is true of great personal discoveries.

Tips for Managing the Station

- Remind students that all surfaces touching bubbles must be wet (hands, straw, table, measuring implements), to dip the straw in solution just prior to blowing, to blow very softly, and not to inhale with their mouths on the straw.

- If you have a student who has great difficulty blowing a table bubble, try blowing a bubble into his soapy hand so he can feel just how softly he needs to blow. Suggest that he begin by blowing a bubble into his own palm prior to blowing a table bubble. If a student takes a quick breath in with her mouth on the straw, she is breaking the soap film at the end of the straw. Cut a diamond-shaped hole in the straw an inch or so below the blowing end to eliminate the problem.

- Keep the table surface covered with a thin coating of bubble solution. Use a squeegee to periodically remove excess bubble solution from the table.

- Throw sections of newspaper over spills on the floor. Refill containers of bubble solution as needed.

Bubble Technology

What to Do

Choose a tool.
Try to make a bubble with it.

Decide whether it makes a big bubble,
a small bubble,
or no bubble at all.

Bubble Technology

Questions

What's
the same about
all the tools that
make bubbles?

Can you turn a non-bubble blower
into a bubble blower?

Activity Task Card for Volunteers

Bubble Technology

Your goal is to assist the students in making their own discoveries, while keeping the activity safe, and the mess under control. Read the signs at the station so you know what the students will be investigating. If the students are non-readers, you will have to communicate the content of each sign. This is best done by giving a challenge or asking a question, rather than demonstrating how it is done. Save your demonstrations for situations when students aren't successful on their own, even with coaching.

Ask the students open-ended questions, such as: What have you discovered? Why do you think that is happening? You may also want to provide further challenges, such as: Can you use that object in a different way to make bubbles?

Resist the temptation to give explanations to the students.

If students get out of control, involved in creating mountains of foam or some other activity that is unrelated to the station, you might want to steer them back on task. Make sure to intervene if you see an unsafe behavior. However, keep in mind that what may appear as fooling around with bubbles can lead to some of the deepest learning experiences. Some of the greatest scientific discoveries have been made while scientists "fooled around"—the same is true of great personal discoveries.

Tips for Managing the Station

- ✌ Discourage students from putting their mouths directly on the objects. Blowing through the object from a distance of several inches or waving the object through the air works more effectively.

- ✌ Foamy soap solution interferes with this activity. Skim the foam off the surface of the bubble solution periodically. You may want to ask your students not to jiggle their objects in the bubble solutions as too much of this is what makes the foam.

- ✌ Refill tubs of bubble solution as needed. There should be enough bubble solution in the tub to completely cover the bubble-making objects.

- ✌ Keep the objects on the table, not in the solution. This will make them easier to find, and the tubs clearer for complete dipping.

- ✌ Use a squeegee to periodically remove excess bubble solution from the table. Throw sections of newspaper over spills on the floor.

Bubble Colors

What to Do

Bubble Colors

Questions

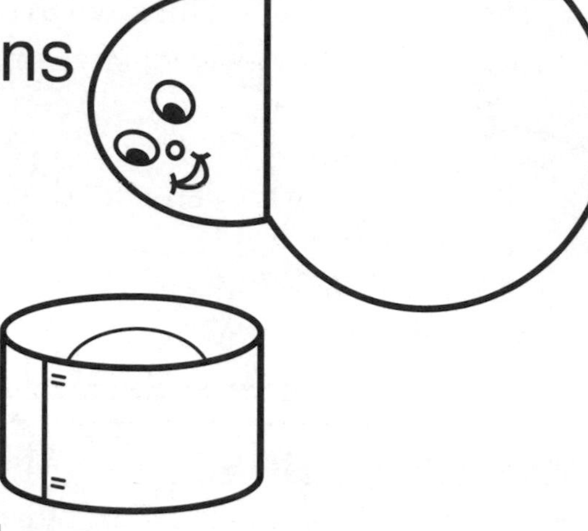

What colors and patterns do you see?

What happens if you blow lightly on a bubble?

Can you tell how old a bubble is by its colors?

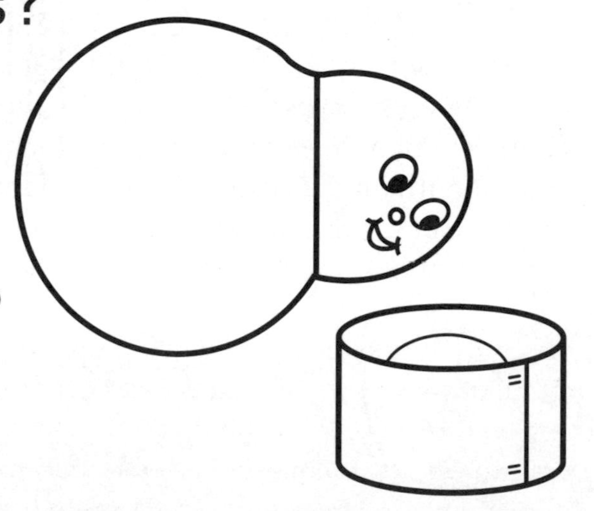

Can you tell exactly when a bubble will pop?

Five, four, three, two, one ...

Activity Task Card for Volunteers

Bubble Colors

Your goal is to assist the students in making their own discoveries, while keeping the activity safe, and the mess under control. Read the signs at the station so you know what the students will be investigating. If the students are non-readers, you will have to communicate the content of each sign. This is best done by giving a challenge or asking a question, rather than demonstrating how it is done. Save your demonstrations for situations when students aren't successful on their own, even with coaching.

Ask the students open-ended questions, such as: What have you discovered? Why do you think that is happening? You may want to provide further challenges, such as: Is there an order to the colored rings? Can you keep a bubble from popping by blowing on it?

Resist the temptation to give explanations to the students.

If students get out of control, you might want to steer them back on task. Make sure to intervene if you see an unsafe behavior. However, keep in mind that what may appear as fooling around with bubbles can lead to some of the deepest learning experiences. Some of the greatest scientific discoveries have been made while scientists "fooled around."

Tips for Managing the Station

- Remind students that all surfaces touching bubbles must be wet (hands, straw, table), to dip the straw in solution just prior to blowing, to blow very softly, and not to inhale with their mouths on the straw.

- If you have a student who has great difficulty blowing a table bubble, try blowing a bubble into his soapy hand so he can feel just how softly he needs to blow. Suggest that he begin by blowing a bubble into his own palm prior to blowing a table bubble. If a student takes a quick breath in with her mouth on the straw, she is breaking the soap film at the end of the straw. Cut a diamond-shaped hole in the straw an inch or so below the blowing end to eliminate the problem.

- This activity works best when small groups blow and observe one bubble at a time until it pops. You may want to organize the students at the station so they take turns being the one to blow the bubble, while the others call out the colors they see. Remind students to use the white bubble "homes." These will reflect more light, allow students to see more colors, and protect the bubble.

- Keep the surface of the black plastic covered with a thin coating of bubble solution. Use a squeegee to periodically remove excess bubble solution from the table. Throw sections of newspaper over spills on the floor, and refill containers of bubble solution as needed.

Bubble Windows

What to Do

Hold on to the straws, and dip the whole loop of string into the solution.

Slowly lift it out!

Bubble Windows

Questions

What happens when you pull your window through the air?

Can you poke things through without popping it?

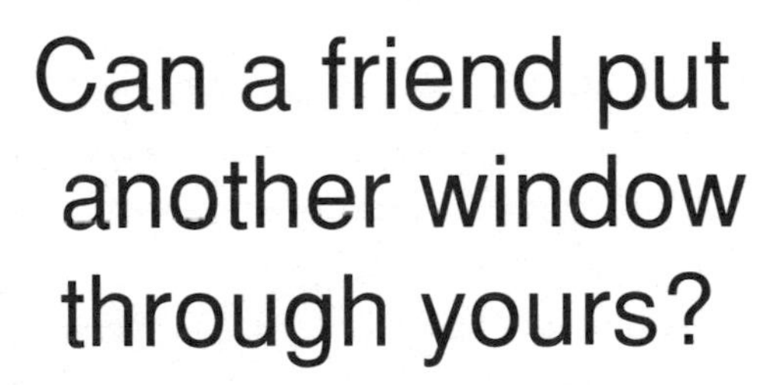

Can a friend put another window through yours?

How many ways can you make a butterfly?

Activity Task Card for Volunteers

Bubble Windows

Your goal is to assist the students in making their own discoveries, while keeping the activity safe, and the mess under control. Read the signs at the station so you know what the students will be investigating. If the students are non-readers, you will have to communicate the content of each sign. This is best done by giving a challenge or asking a question, rather than demonstrating how it is done. Save your demonstrations for situations when students aren't successful on their own, even with coaching.

Ask the students open-ended questions, such as: What have you discovered? Why do you think that is happening? You may also want to provide further challenges, such as: Hold a bubble window vertically for a minute or two. Can you notice a pattern to the colors you see?

Resist the temptation to give explanations to the students.

If students get out of control, involved in creating mountains of foam or some other activity that is unrelated to the station, you might want to steer them back on task. Make sure to intervene if you see an unsafe behavior. However, keep in mind that what may appear as fooling around with bubbles can lead to some of the deepest learning experiences. Some of the greatest scientific discoveries have been made while scientists "fooled around."

Tips for Managing the Station

- If students are having difficulty making bubble windows, remind them that all surfaces touching bubbles must be wet (hands, strings, and straws).

- Suggest that students do everything in slow motion at this station. This will help them be more successful, and it nicely moderates the excitement level.

- Lifting the string and straw out of the bubble solution at an angle, rather than straight up, is more effective.

- Discourage students from walking blindly backwards to make bubbles. Encourage them to wave their windows through the air using an up-down motion while standing still instead.

- Foamy soap solution interferes with this activity. Skim the foam off the surface of the bubble solution periodically. You may want to ask your students not to jiggle their windows in the bubble solution as too much of this makes foam.

- Refill tubs of bubble solution as needed. There should be enough bubble solution in the tub to completely cover the string-and-straws frame.

- Use a squeegee to periodically remove excess bubble solution from the table. Throw sections of newspaper over spills on the floor.

Bubble Walls

What to Do

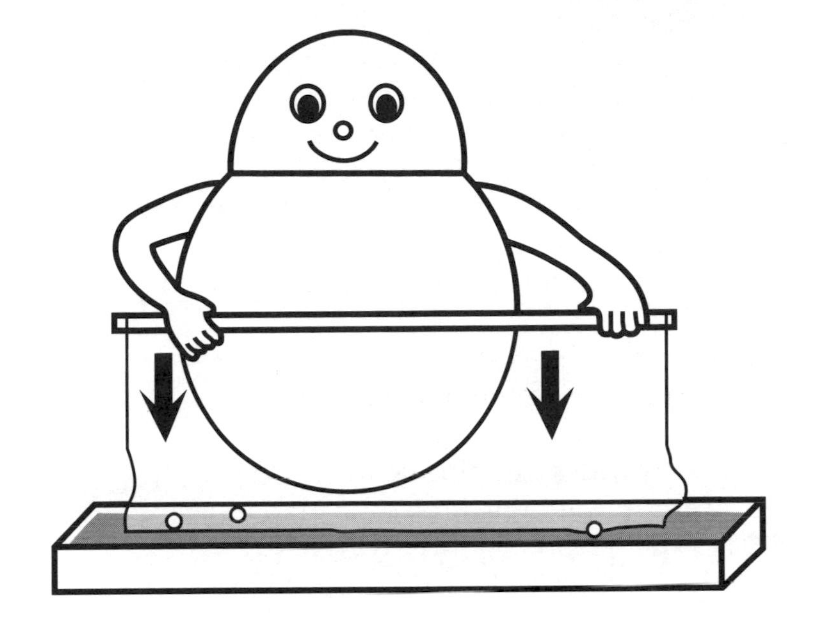

Dip the stick and string into the solution …

… and slowly lift the stick out!

Bubble Walls

Questions

What do things look like through the bubble wall?

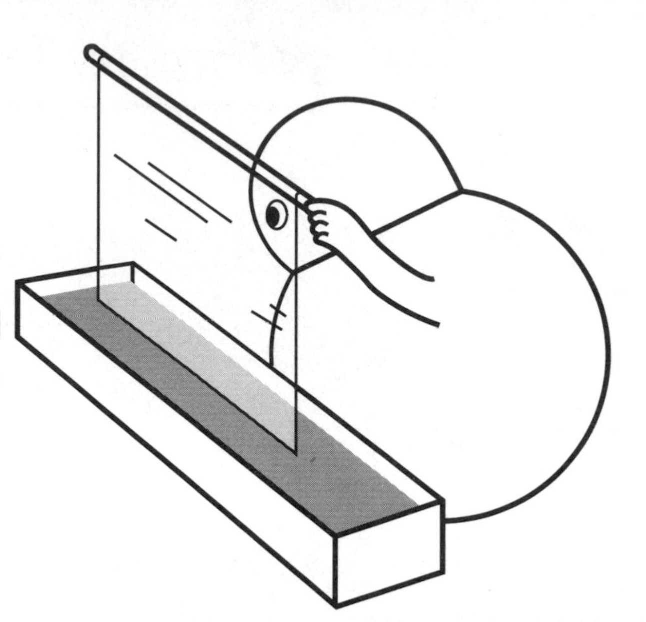

Can you make the wall sway?

Can you shake hands through the window?

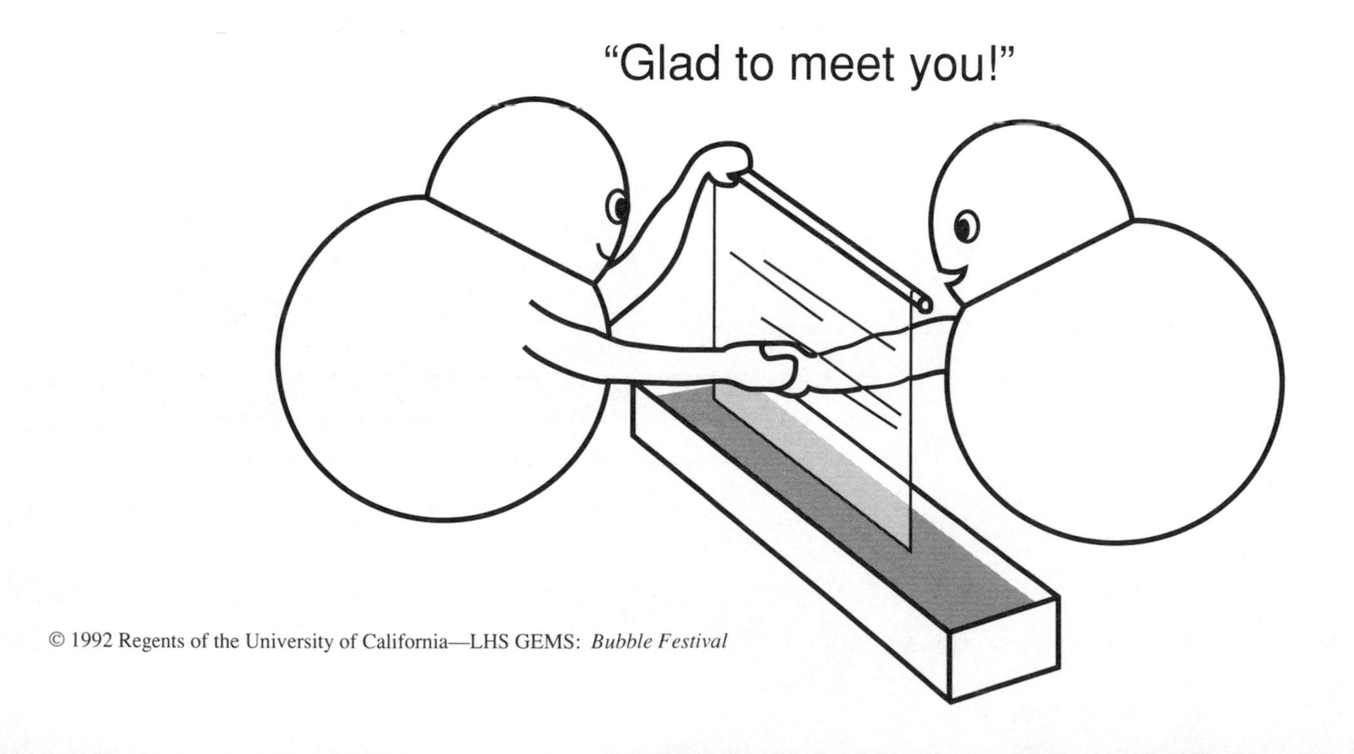

Activity Task Card for Volunteers

Bubble Walls

Your goal is to assist the students in making their own discoveries, while keeping the activity safe, and the mess under control. Read the signs at the station so you know what the students will be investigating. If the students are non-readers, you will have to communicate the content of each sign. This is best done by giving a challenge or asking a question, rather than demonstrating how it is done. Save your demonstrations for situations when students aren't successful on their own, even with coaching.

Ask the students open-ended questions, such as: What have you discovered? Why do you think that is happening? You may also want to provide further challenges, such as: Can you blow a small bubble from a bubble wall? Can you make a bubble wall wiggle?

Resist the temptation to give explanations to the students.

If students get out of control, involved in creating mountains of foam or some other activity that is unrelated to the station, you might want to steer them back on task. Make sure to intervene if you see an unsafe behavior. However, keep in mind that what may appear as fooling around with bubbles can lead to some of the deepest learning experiences. Some of the greatest scientific discoveries have been made while scientists "fooled around."

Tips for Managing the Station

- ✌ If students are having difficulty making bubble walls, mention that the string has to be inside the trough. Also remind them that all surfaces touching bubbles must be wet (hands, strings, and dowels).

- ✌ Suggest that students do everything in slow motion at this station. This will help them be more successful, and it nicely moderates the excitement level.

- ✌ Foamy soap solution interferes with this activity. Skim the foam off the surface of the bubble solution periodically. You may want to ask your students not to jiggle the walls in the bubble solutions as too much of this makes foam.

- ✌ Refill tubs of bubble solution as needed. There should be enough bubble solution in the tub to completely cover the dowel-and-string frame.

- ✌ Use a squeegee to periodically remove excess bubble solution from the table. Throw sections of newspaper over spills on the floor.

Bubble Foam

What to Do

Make lots of foam!

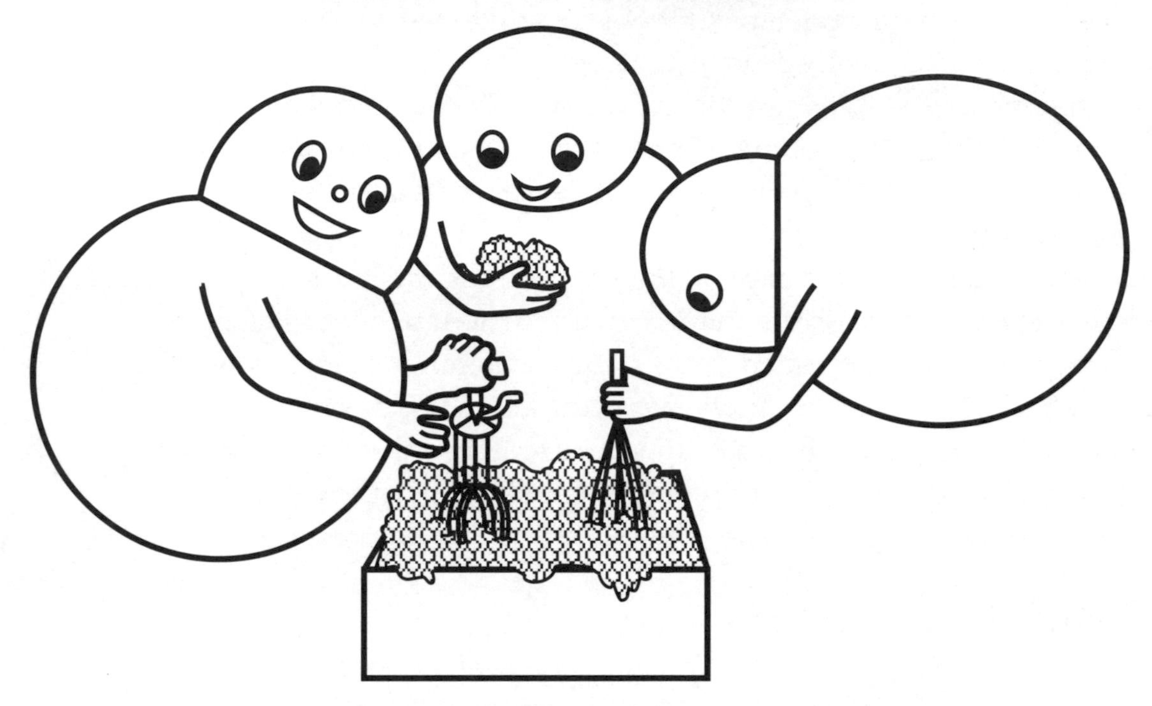

Can you fill a bucket with foam?

Bubble Foam

Questions

What is foam?

Can you count the tiny bubbles?

Put your straw in
and blow.

Is this foam, too?

Activity Task Card for Volunteers

Bubble Foam

Your goal is to assist the students in making their own discoveries, while keeping the activity safe, and the mess under control. Read the signs at the station so you know what the students will be investigating. If the students are non-readers, you will have to communicate the content of each sign. This is best done by giving a challenge or asking a question, rather than demonstrating how it is done. Save your demonstrations for situations when students aren't successful on their own, even with coaching.

Ask the students open-ended questions, such as: What have you discovered? Why do you think that is happening? You may also want to provide further challenges, such as: Can you make foam with smaller bubbles? With bigger bubbles? How does the foam change if you beat faster? Can you count the bubbles?

Resist the temptation to give explanations to the students.

If students get out of control, involved in using egg beaters to make music or some other activity that is unrelated to the station, you might want to steer them back on task. Make sure to intervene if you see an unsafe behavior. However, keep in mind that what may appear as "fooling around" with bubbles can lead to some of the deepest learning experiences. Some of the greatest scientific discoveries have been made while scientists "fooled around" — the same is true of great personal discoveries.

Tips for Managing the Station

- ✌ Keep your eye on the transfer of foam to bucket. If students are not doing this themselves, then you will want to periodically remove foam from the solution by scooping into the bucket. If the bucket gets too full, dump foam in a nearby, plastic-lined garbage can. After several hours, the foam will "disappear."

- ✌ Don't allow the students to put foam on their faces.

- ✌ Refill tubs of bubble solution as needed.

- ✌ Use a squeegee to periodically remove excess bubble solution from the table.

- ✌ Throw sections of newspaper over spills on the floor.

Bubble Skeletons

What to Do

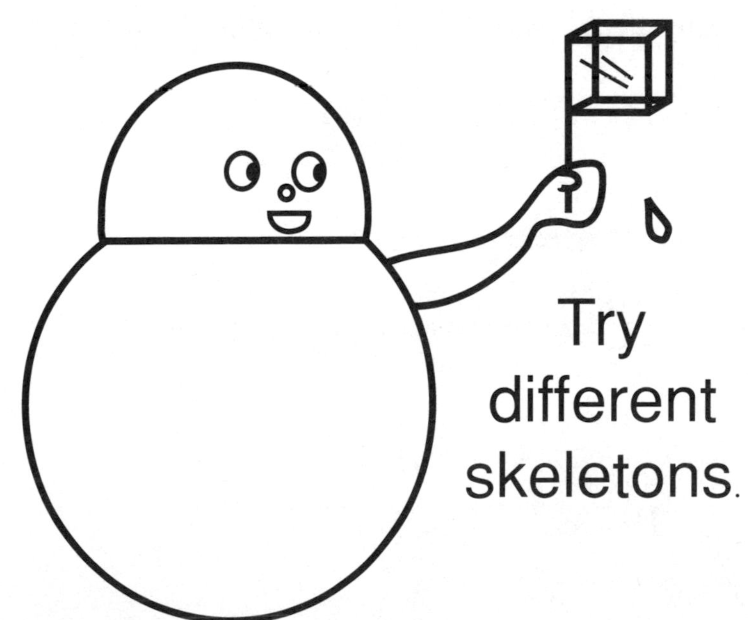

Bubble Skeletons

Questions

Try popping one side at a time.
How does the soap film change?

Can you make a little square bubble inside the cube skeleton? (Try double-dipping!)

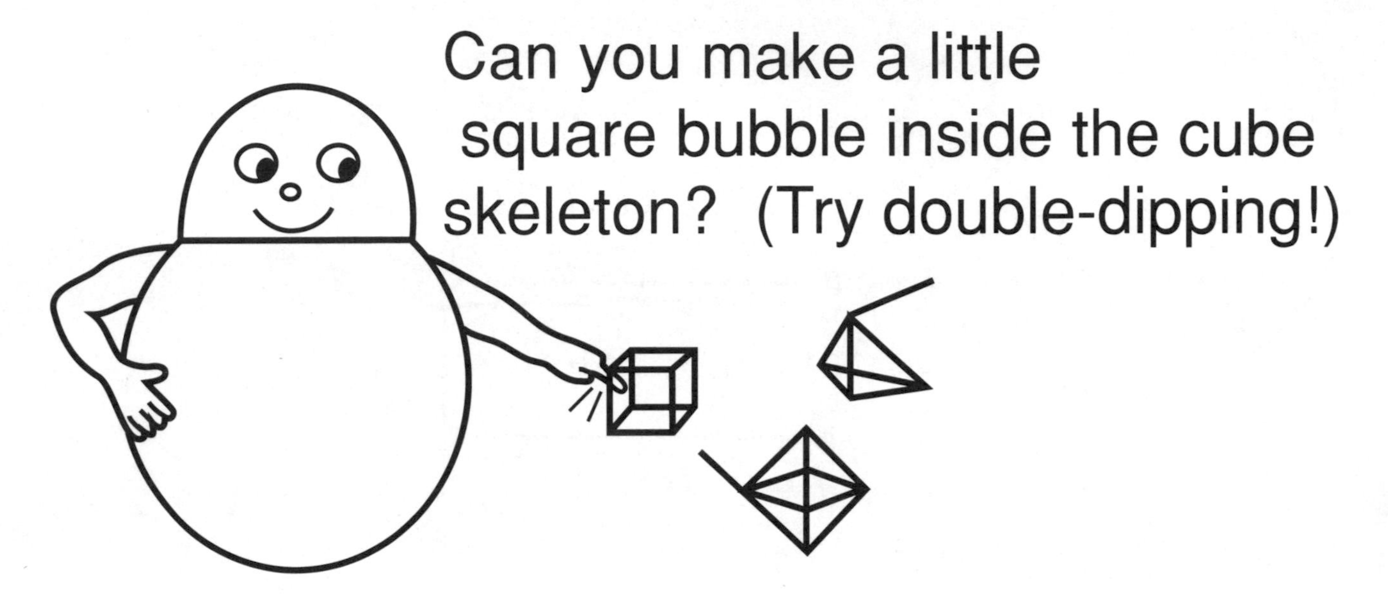

Wave the skeletons through the air or blow into them with a straw.

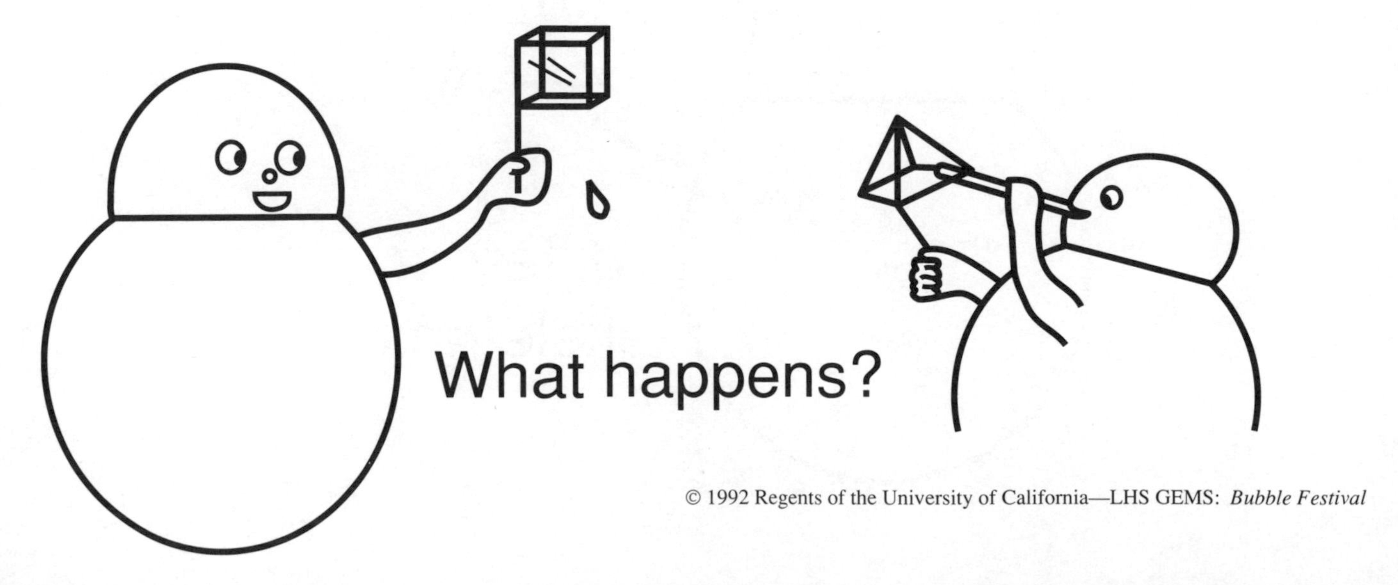

What happens?

Activity Task Card for Volunteers

Bubble Skeletons

Your goal is to assist the students in making their own discoveries, while keeping the activity safe, and the mess under control. Read the signs at the station so you know what the students will be investigating. If the students are non-readers, you will have to communicate the content of each sign. This is best done by giving a challenge or asking a question, rather than demonstrating how it is done. Save your demonstrations for situations when students aren't successful on their own, even with coaching.

Ask the students open-ended questions, such as: What have you discovered? Why do you think that is happening? You may also want to provide further challenges, such as: Can you predict how the soap film will change when you pop one side? Why do you suppose it changes in that way?

Resist the temptation to give explanations to the students.

If students get out of control, involved in waving the skeletons too wildly or some other
activity that is unrelated to the station, you might want to steer them back on task. Make sure to intervene if you see an unsafe behavior. However, keep in mind that what may appear as "fooling around" with bubbles can lead to some of the deepest learning experiences. Some of the greatest scientific discoveries have been made while scientists "fooled around" — the same is true of great personal discoveries.

Tips for Managing the Station

- Remind students that all surfaces touching bubbles must be wet (hands, arms, etc.).

- Use a squeegee to periodically remove excess bubble solution from the table.

- Throw sections of newspaper over spills on the floor.

- Refill containers of bubble solution as needed. The solution in the dish pans should be high enough to cover the skeletons.

Frozen Bubbles

What to Do

Blow a bubble over the bucket.
Watch as it falls into the bucket.

It works better
if you don't blow
down into the bucket.

NEVER TOUCH
THE DRY ICE!

Frozen Bubbles

Questions

What happened to the bubble?
Where did it go?

Did the bubble change?
How?

Activity Task Card for Volunteers

Frozen Bubbles

Your goal is to assist the students in making their own discoveries, while keeping the activity safe, and the mess under control. Read the signs at the station so you know what the students will be investigating. If the students are non-readers, you will have to communicate the content of each sign. This is best done by giving a challenge or asking a question, rather than demonstrating how it is done. Save your demonstrations for situations when students aren't successful on their own, even with coaching.

Ask the students open-ended questions, such as: What have you discovered? Why do you think that is happening? You may also want to provide further challenges, such as: What is happening to the bubbles as time goes by? How are they changing?

Resist the temptation to give explanations to the students.

This is a station for quiet contemplation and discussion. Direct students to blow bubbles above the bucket and let them fall in. By blowing downward into the bucket, they may disrupt the layer of carbon dioxide gas that forms near the dry ice. **Be sure that no one puts their hands into the bucket.** Dry ice (frozen carbon dioxide) is so cold that it can "burn" the skin. Make sure to intervene if you see an unsafe behavior.

Tips for Managing the Station

- Use gloves when working with or near the dry ice.
- If a student blows into the bucket and disrupts the layer of carbon dioxide, just wait. The layer will re-form.
- Throw sections of newspaper over spills on the floor.
- Refill containers of bubble solution as needed.

Stacking Bubbles

What to Do

Put your straw into the rubber hose and blow.

Watch the bubbles
in between the clear,
plastic walls.

Stacking Bubbles

Questions

What shape are the bubbles when they are stacked?

How many sides do most bubbles have?

Do the sides and corners of the bubbles join in any special way?

Activity Task Card for Volunteers

Stacking Bubbles

Your goal is to assist the students in making their own discoveries, while keeping the activity safe, and the mess under control. Read the signs at the station so you know what the students will be investigating. If the students are non-readers, you will have to communicate the content of each sign. This is best done by giving a challenge or asking a question, rather than demonstrating how it is done. Save your demonstrations for situations when students aren't successful on their own, even with coaching.

Ask the students open-ended questions, such as: What have you discovered? Why do you think that is happening? You may also want to provide further challenges, such as: Count the sides and "corners" of some of the bubbles. Do you see a pattern in the way the bubbles join together? Do the bubble patterns remind you of other things you've seen?

Resist the temptation to give explanations to the students.

If students get out of control, involved in creating mountains of foam or some other activity that is unrelated to the station, you might want to steer them back on task. Check to be sure that students are using their straws to blow, rather than blowing directly into the rubber tubing. Make sure to intervene if you see an unsafe behavior.

Tips for Managing the Station

- Check periodically to see that one end of the rubber tubing is placed between the sheets of plastic and **in** the bubble solution.

- Suggest that students blow very softly at this station. This will help them be more successful.

- Foamy soap solution interferes with this activity. Skim the foam off the surface of the bubble solution periodically.

- Refill the tub of bubble solution as needed. There should be enough bubble solution in the tub to cover the bottom edges of the sheets of clear plastic and the end of the rubber tubing.

- Use a squeegee to periodically remove excess bubble solution from the table.

- Throw sections of newspaper over spills on the floor.

Swimming Pool Bubbles

What to Do

Get in line for Swimming Pool Bubbles.

When it's your turn, step carefully onto the box.

Be sure to keep your hands down and your eyes open!

Swimming Pool Bubbles

Questions

What is it like to be inside a bubble?

Can you be a bubble buddy?

Activity Task Card for Volunteers

Swimming Pool Bubbles

Your goal is to assist the students in making their own discoveries, while keeping the activity safe, and the mess under control. Read the signs at the station so you know what the students will be investigating. If the students are non-readers, you will have to communicate the content of each sign. This is best done by giving a challenge or asking a question, rather than demonstrating how it is done. Save your demonstrations for situations when students aren't successful on their own, even with coaching.

Ask the students open-ended questions, such as: What is it like to be inside a bubble? What do you notice about the colors on the bubble film? You may also want to provide further challenges, such as: Why do you suppose the bubble film seems to pull inward?

Resist the temptation to give explanations to the students.

Students who are waiting their turn should stand in line and observe and discuss what they see. Help each student on and off the platform, reminding them to step carefully. Make sure to intervene if you see an unsafe behavior.

Tips for Managing the Station

- Before you begin, tell the students to step on and off the crate *slowly*.
- Keep the surface of the crate as dry as possible.
- Be sure that all surfaces that may touch the bubble are wet, including your hands and the inner sides of the wading pool.
- Remind students to hold in their arms and any bulky clothing.
- To avoid back strain as you bend to lift the hula hoop up around each child, you may want to kneel, rather than stand.
- Use a squeegee or a towel to remove excess bubble solution from the floor around the wading pool, and throw sections of newspaper over wet areas, especially where students mount and dismount from the pool.